ADVANCES IN

NANODEVICES AND NANOFABRICATION

edited by | Qing Zhang | W. I. Milne | Jianping Zou

ADVANCES IN NANODEVICES AND NANOFABRICATION

Selected Publications from Symposium of Nanodevices and Nanofabrication in ICMAT2011

PAN STANFORD PUBLISHING

Published by

Pan Stanford Publishing Pte. Ltd.
Penthouse Level, Suntec Tower 3
8 Temasek Boulevard
Singapore 038988

Email: editorial@panstanford.com
Web: www.panstanford.com

British Library Cataloguing-in-Publication Data
A catalogue record for this book is available from the British Library.

Advances in Nanodevices and Nanofabrication:
Selected Publications from Symposium of Nanodevices and Nanofabrication in ICMAT2011

ISBN 978-981-4364-54-6 (Hardcover)
ISBN 978-981-4364-55-3 (eBook)

Contents

Preface xi

1. Fabrication of Nanowires for Biosensing Applications 1

Thomas Moh Shan Yau, Gregory Pandraud, Louis C.P.M. de Smet, Ernst J.R. Sudhölter, Cees J.M. van Rijn, and Pasqualina M. Sarro

1.1 Introduction 2
1.2 The Growth of Nanowires (Bottom-Up Approaches) 4
1.2.1 Vapor–Liquid–Solid 5
1.2.2 Non-Catalytic Growth: Vapor Solid (VS) 7
1.2.3 Template-Based Growth 8
1.2.4 Growing via Solution 10
1.2.5 Orderly Growing: A Hybrid Approach 10
1.3 Top-Down Approaches 14
1.3.1 Beams Lithography 14
1.3.2 Spacer/Edge Transfer Lithography 16
1.3.3 Nanoimprint Technology 17
1.3.4 Oxidation 18
1.3.5 Deposition and Etching under Angle 20
1.3.6 Anisotropic Etched Microfabrication SiNWs 22
1.3.7 SiNW Field Effect Transistor 26
1.4 Top-Down Versus Bottom-Up Approaches 29
1.5 Conclusion and Future Challenges 31

2. Novel Nanostructured Carbon Nanotube Electron Sources 41

Matthew T. Cole, Mark Mann, Chi Li, Kai Hou, Yan Zhang, Xe Qing, Kenneth B. K. Teo, Husnu Emrah Unalan, Wei Lei, Baoping Wang, Daping Chu, Didier Pribat, Pierre Legagneux, Laurent Gangloff, Vincent Semet, Pritesh Hiralal, Gehan A. J. Amaratunga, and William I. Milne

2.1 Electron Emission 42
2.1.1 The Field Emission Mechanism and the Fowler–Nordheim Equation 42

2.1.2 State-of-the-Art Electron Emitters 44
2.1.3 The History of Carbon-Based Field Emitters 49
2.1.4 Carbon Nanotube and Nanofiber-Based Field Emitters 50
2.2 CNT Thin Films on Flexible Substrates 52
2.2.1 CNT Thin Film Deposition 52
2.2.2 Ballasted CNT/ZnO Nanowire Thin Films 55
2.3 Periodic Field Emitters 58
2.3.1 Periodic Arrays 58
2.3.2 SOI Substrate Ballasted CNT Arrays 59
2.3.3 ZnO NW Ballasted CNT Arrays 62
2.3.4 Hexagonal Mesh Arrays 64
2.4 CNT Deposition on Conducting Substrates 65
2.4.1 Stainless Steel Mesh 66
2.4.2 W Tips 68
2.5 Summary 70

3. Plasma-Enhanced Chemical Vapor Deposition of Graphene Nanostructures 75
Timothy van der Laan, Shailesh Kumar, and Kostya (Ken) Ostrikov
3.1 Introduction 75
3.2 Successful Approaches to Synthesizing Graphene 76
3.2.1 Mechanical Exfoliation 77
3.2.2 Chemical Synthesis 78
3.2.3 Supported Growth 79
3.2.4 Summary of Graphene Synthesis 80
3.3 Advantages of Plasma-Enhanced Chemical Vapor Deposition Approaches 81
3.3.1 Plasma-Related Effects 81
3.3.1.1 Building units 81
3.3.1.2 Plasma heating 82
3.3.1.3 Surface activation 83
3.3.1.4 Usefulness of plasmas 84
3.3.2 Plasma Enhancement Successful for Other Carbon Structures 86
3.3.2.1 Carbon nanotubes 86
3.3.2.2 Vertically aligned graphene nanosheets and carbon nanowalls 87

3.4 Current Status of PECVD of Graphene 88
3.4.1 General Review 88
3.4.2 Authors' Recent Contributions 91
3.5 Concluding Remarks 92

4. Morphology Control at the Nanoscale of Polymer Composites for Organic Solar Cells 99

Bernard Ratier, Johann Bouclé, Ghislain Boloma, Fabrice Rossignol, and Valérie Coudert

4.1 Introduction 100
4.2 P3HT-PCBM-Based Solar Cells 103
4.2.1 Main Photovoltaic Parameters 103
4.2.2 Correlation Between Device Processing and Efficiency 105
4.2.3 Nanoscale Characterization of the Blend Structure and Carrier Photogeneration Using Near-Field Microscopy 108
4.3 Conclusion 112

5. Theoretical Studies on Architectures of Straight Zigzag//Armchair Carbon Nanotube Junctions as Molecular Electronic Devices 117

Anurak Udomvecha and Teerakiat Kerdcharoen

5.1 Introduction 118
5.2 Modeling and Computational Procedures 119
5.3 Results and Discussion 120
5.4 Conclusion 127

6. Energy Dissipation Analysis of Defected Carbon Nanotube Oscillators 129

Taiyu Guo, Tony Weixi Ding, Qing-Xiang Pei, and Yong-Wei Zhang

6.1 Introduction 130
6.2 Simulation Methodology 130
6.3 Results and Discussion 131
6.3.1 Effect of sp^3-Hybridized Defects on Carbon Nanotube Oscillators 131
6.3.2 Mechanisms of Energy Dissipation 135
6.4 Conclusion 135

7. **Single-Wall Carbon Nanotube-Based Transparent and Conductive Films** **137**

Zhiyu Yan, Xuena Luan, Yu Fan, and Qing Zhang

7.1 Introduction 137
7.2 Experimental Procedure 139
7.3 Results and Discussion 140
7.3.1 The Effects of Number of Spin Coatings 140
7.3.2 The Effects of Centrifugation 140
7.3.3 The Effects of APTES Treatment 141
7.3.4 The Effects of Soft Baking 142
7.3.5 The Effects of Hard Baking 144
7.3.6 The Effects of Acid Treatment 144
7.3.7 The Effects of PEDOT Treatment 145
7.3.8 The Effects of the Combination of Acid and PEDOT Treatment 147
7.4 Conclusions 148

8. **Effect of Edge Passivated by Hydrogen on the Transport Properties of Finite-Size Metallic Carbon Nanotube-Based Molecular Devices** **153**

Zongling Ding, Jun Jiang, Huaizhong Xing, Haibo Shu, Yan Huang, Xiaoshuang Chen, and Wei Lu

8.1 Introduction 154
8.2 Model and Method 155
8.3 Results and Discussions 156
8.4 Conclusion 159

9. **A Study of Defect-Induced Electrical Breakdown on Individual Multi-Wall Carbon Nanotubes** **163**

Gopal Krishna Goswami and Karuna Kar Nanda

9.1 Introduction 164
9.2 Experiment 165
9.3 Results and Discussions 166
9.4 Conclusions 169

10. Carbon Nanotube- and Graphene-Based Micro-Sensors and Reactors **171**

Raluca Savu, Rogério Valentim Gelamo, Francisco Paulo Marques Rouxinol, Alfredo Rodrigues Vaz, Gustavo Monteiro da Silva Valente, Alexander Flacker, Mara Adriana Canesqui, Angelo Luiz Gobbi, and Stanislav Moshkalev

10.1 Introduction 171
10.2 Experimental Procedure 173
10.3 Results and Discussion 175

11. Carrier Statistics Modeling of p-Type Graphene Nanoribbons **179**

Seyed Mahdi Mousavi, Mohammad Taghi Ahmadi, Jeffrey Frank Webb, Zaharah Johari, Azizah Amin, and Razali Ismail

11.1 Introduction 180
11.2 Graphene Nanoribbon Band Structure 181
11.3 Equilibrium Carrier Statistics 184
11.4 Conclusion 186

12. Various Tunable Photonic Crystal/Quasi-Photonic Crystal Structures Fabricated by Reconfigurable Interference System **189**

Haitao Dai, Xiaowei Sun, Yanjun Liu, and Dan Luo

12.1 Introduction 190
12.2 Methodology 191
12.3 Theoretical Description 192
12.4 Experimental Results 193
12.5 Further Analysis 195
12.6 Conclusion 197

13. Silicon Waveguide-Based Polarization Rotator **201**

Huijuan Zhang, Jing Zhang, Shiyi Chen, Liang Ding, Tsung-Yang Liow, Mingbin Yu, and Guo-Qiang Lo

13.1 Introduction 202
13.2 Design and Simulation 203
13.3 Fabrication Methods 206

13.4 Results and Discussion 207
13.5 Conclusions 210

14. Silicon Microring Resonator-Based Thermo-Optic Wavelength Selective Switch for Optical Communication Network **213**
Prabhathan Patinharekandy, Jing Zhang, Huijuan Zhang, Guo-Qiang Lo, and Murukeshan Vadakke Matham
14.1 Introduction 214
14.2 Device Design and Fabrication 215
14.3 Experimental Results and Discussion 217
14.3.1 Discrete and Fine Wavelength Tuning 217
14.3.2 Wavelength Selection in a Broadband DWDM System 219

15. Tuning Plasmonic Resonance Wavelength by Ordered Silver Hole Arrays on Si and SiO_2 **221**
Hongwei Gao, Benzhong Wang, Jinghua Teng, Soo Jin Chua, and Ning Xiang
15.1 Introduction 222
15.2 Fabrication 223
15.3 Results and Discussion 224
15.4 Conclusion 229

16. Revelation of Temperature-Dependent Cathodoluminescence by Employing Nanostructural Pathway in Fe-Incorporated ZnO **231**
Jai Shankar Tawale, Hirotsugu Nishido, Shin Toyoda, Melepurath Deepa, Nupur Bahadur, Mahalingam Maniraj, Sudipta Roy Barman, Bibhash Ranjan Chakraborty, and Avanish Kumar Srivastava
16.1 Introduction 231
16.2 Experimental 233
16.3 Results and Discussion 234
16.3.1 Growth Morphologies and Phase Formations 234
16.3.2 Spectroscopy and Emission Bands 240
16.4 Conclusions 247

17. Raman and Photoemission Spectroscopy Studies on Thermal Stability of Quantum Dot Consisting of Si Clad and Ge Core **251**

Yudi Darma

17.1 Introduction 252
17.2 Experimental Detail 252
17.3 Result and Discussion 253
17.4 Conclusions 256

18. Application of TEM Nano-Beam Diffraction Strain Analysis in Advanced Semiconductor Devices **259**

Zhu Jie, Zhou Yong Kai, Du An Yan, and Hua You Nan

18.1 Introduction 259
18.2 Experiment 262
18.2.1 Sample Preparation 262
18.2.2 NBED Analysis 262
18.3 Results and Discussion 264
18.3.1 Thermal Stress-Induced Si Defects 264
18.3.2 Thermal Stress-Induced Si Defects 268
18.4 Conclusion 270

19. Electron Mobility and Persistent Photoconductivity in Quantum Wells $In_{0.52}Al_{0.48}As/In_{0.53}Ga_{0.47}As/In_{0.52}Al_{0.48}As$ on InP Substrate **273**

Vladimir A. Kulbachinskii, Roman A. Lunin, Natalia A. Yuzeeva, Galib B. Galiev, Ivan S. Vasilievskii, and Eugene A. Klimov

19.1 Introduction 274
19.2 Experimental 274
19.3 Results and Discussion 275
19.4 Conclusion 281

Index 283

Preface

This book provides a selection of manuscripts submitted to the 4th International Symposium of Nanodevices and Nanofabrication, one of 37 symposia held as part of the 6th International Conference on Materials for Advanced Technologies (ICMAT 2011), which took place at the Suntec Singapore International Convention and Exhibition Centre, Singapore, from June 26 to July 1, 2011.

"Nanoscience and nanotechnology" have become an indispensable multidisciplinary research area. A variety of devices at the nanometer scale/molecular scale for electronic, biological, and mechanical applications have been created through the rapid development of materials and fabricating technologies. This symposium served as a forum for highlighting advances in nanoelectronic devices, nanomechanics and NEMS, the synthesis and characterization of nanostructures, the fabrication, manipulation, and aligning processes at the nanometer scale, quantum phenomena, theoretical modeling of nanodevices and nanomaterials, etc.

The symposium had excellent support, with 20 invited talks, 46 oral presentations, and 48 posters. Results were presented by participants from 15 countries and regions.

We would like to express our gratitude to all delegates for their active participation in the symposium's activities.

Qing Zhang
Chair of the Symposium
School of EEE, Nanyang Technological University
Singapore

William I. Milne
Co-Chair of the Symposium
Department of Electrical Engineering
University of Cambridge
United Kingdom

Jianping Zou
School of EEE, Nanyang Technological University
Singapore

Chapter 1

Fabrication of Nanowires for Biosensing Applications

Thomas Moh Shan Yau,[a,b,*] Gregory Pandraud,[a] Louis C.P.M. de Smet,[b] Ernst J.R. Sudhölter,[b] Cees J.M. van Rijn,[c] and Pasqualina M. Sarro[a]

Delft University of Technology, The Netherlands, [a] Delft Institute of Microsystems and Nanoelectronics, [b] Department of Chemical Engineering, Nano Organic Chemistry, [c] Nanosens, The Netherlands

*t.s.y.moh@tudelft.nl

The fabrication or synthesis of nanowires (NWs) has been studied intensively over the past decades. The unique physical properties of NWs and the ability to offer highly sensitive detection make NW-based biosensors a promising choice for next generation biosensors, especially for medical, food and environment applications. Deeper and thorough understanding on how to grow or fabricate NWs for biosensing purposes is crucial. This chapter discusses various techniques, either with bottom-up or top-down approaches, to synthesize or fabricate NWs. The advantages and disadvantages of bottom-up and top-down methods will also be discussed and weighted, thus providing an idea of the more suitable or promising techniques for making SiNW FETs (field effect transistor) as sensors. The importance of the field effect and doping concentrations inside

Advances in Nanodevices and Nanofabrication: Selected Publications from Symposium of Nanodevices and Nanofabrication in ICMAT2011
Edited by Qing Zhang, W. I. Milne, and Jianping Zou

ISBN 978-981-4364-54-6 (Hardcover), 978-981-4364-55-3 (eBook)
www.panstanford.com

NWs for biosensing devices are summarized and compared. Finally, issues and future challenges related to SiNW FETs as biosensors will also be briefly discussed.

1.1 Introduction

The unwanted discovery of metal whiskers that cause short circuit and arching phenomena in the electrical equipments belonging to Bell Telephone Company back in the 1940s and 1950s unexpectedly triggered the start of synthesizing microsize or nanosize wires in laboratory for research purposes [1]. The term "whiskers," which is used in the animal kingdom describing the usage of hairs as tactile sensation, was first coined to represent microsize wires synthesized in laboratories before the microwire and subsequently the NW terms were used. Since then, tremendous research efforts have been made to study NWs especially in nanoelectronics, bio-sensing and nanophotonics [2–4]. The "center of dogma" or the "holy grail" of using NWs as material for biosensing actually lies in the ability of NWs to provide label free, fast and real time detection, besides offering ultrahigh sensitivity detection for small concentration of a sample, which is not available in current sensing apparatus. Understanding what defines a NW is very important in order to capture the essence of this technology.

NW-based sensing devices have been used to detect viruses [5], proteins [6, 7], DNAs [8–10], cell culture monitoring [11] and as immunosensors [12]. Understanding the working principles of NW-based sensors is important to choose suitable fabrication techniques.

Binding of specific species or ligands onto the chemically modified surface of a NW will alter the electrical charges inside the NW, thus changing its conductance [5]. However, transducing and translating the signal change can be divided into two schemes: the resistivity and the FET scheme, as schematically depicted in Figure 1.1. The resistivity scheme applies mostly to the metallic NW-based sensors while the FET scheme is used in semiconductor NW-based sensing devices. In the resistivity scheme, the metallic NWs act as metal resistors and any changes (binding) on their surface will alter the conductance inside the NWs. This change will increase or decrease the signal relatively to the previously measured signal. Therefore, a signal change indicates the present of targeted species. No extra biasing is required in operating these devices.

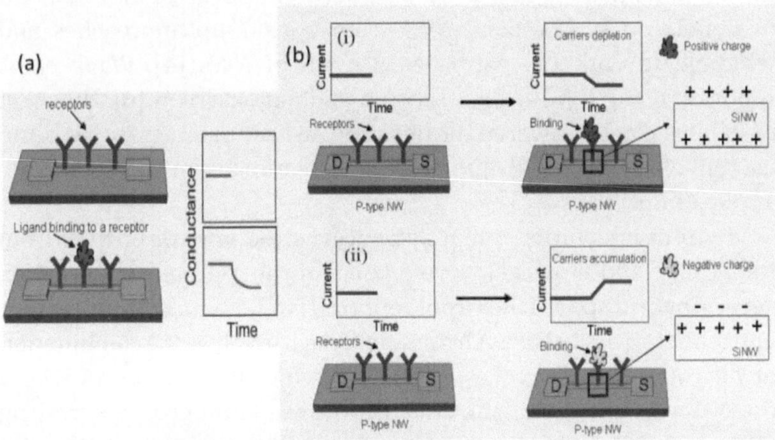

Figure 1.1 Metallic resistivity NWs and semiconductor FET NW sensing principles. (a) Resistivity NW: A ligand binding to a receptor will change the conductance inside the NW and no biasing required. (b) A P-type (positive charge) SiNW FET device being biased to operate at the optimum sensitivity regime: (i) A positive charge ligand will deplete the NW (positive charge) resulting in a decrease in the measured signal; (ii) A negative charge ligand will accumulate the NW and an increase in is signal observed. D = drain and S = source (adapted from refs. [5, 13]).

In the FET schemes, NW nanosensors are biased and operated in the highest sensitivity regime. NW FET devices have been reported to operate in different backgate voltage biasing configurations, including negative, zero and positive voltages [13]. The highest sensitivity regime for a low frequency and small signal detection (for biosensing) is the region between accumulation (collection of same charges) and depletion (repulsion between different charges) as explained by Hien *et al.* [13]. Similar to a conventional MOSFET, this region is generated when a field effect phenomenon is observed inside the NWs or simply at the threshold voltage of the NW FET-based devices [12–14]. Compared with metallic resistor NW scheme (no biasing), the ability to bias the NW FET devices to a correct operating regime will increase the sensitivity of the fabricated devices, as illustrated in Figure 1.1.

Two well-known fabrication approaches have been outlined in this field, namely top-down and bottom-up. Fan *et al.* discussed in details on synthesizing NWs from bottom-up approaches and winding up with the patterned growth of NWs [4]. Wang *et al.* reported using NWs as nanosensors and nanocarriers [6]. Huang *et al.* explored and answered the question on how to integrate NWs into current electronics and optoelectronics platforms for more versatile applications [15].

Bottom-up synthesis mainly concerns the growth of NWs from the atomic and molecular level via chemical approaches. However, several hybrid approaches were reported lately, taking the bottom-up approach to a new phase. A hybrid approach involves the combination of microfabrication process steps to define the position of where the NW will grow, thus introducing localized and ordered growing [16]. Although bottom-up approaches could synthesize NWs with dopants such as boron and phosphorous while growing [17], most of the bottom-up approaches produce metallic type NWs. On the other hand, the top-down approach, which is taking full advantage of the mature microfabrication technology for integrated circuits (IC) fabrication, produces FET type NWs. Parameters control of doping in NWs such as concentration, depth, etc., are easier in the conventional IC fabrication processes. Most of the processes are compatible with batch manufacturing, thus promoting mass production. This paper aims at a general discussion on NW fabrication techniques either from bottom-up and top-down approaches, which are suitable for biosensing purposes.

1.2 The Growth of Nanowires (Bottom-Up Approaches)

Inspired by Mother Nature, the bottom-up approach refers to the art of building a structure starting from the very basic element and continuously moving upwards to form a complex and functional structure. Taking biology as example, where a group of similar cells will form tissue and similar functional type of tissues will form an organ and several organs will be integrated and form a complex system such as digestion system and subsequently the whole anatomy of human/animal body. Similar concepts could be adapted in understanding the growth of NWs. Chemically synthesized NWs or growing the NWs by continuously adding atomic or molecular species of a specific material to form more complex structures has been widely reported [4, 15, 18]. In this section, the most important paradigms of growing NWs such as vapor–liquid–solid (VLS), vapor–solid (VS) and template-based electrochemical deposition will be discusses in detail.

1.2.1 Vapor–Liquid–Solid

The growing of a crystalline material from a supersaturated solution through a precipitation process serves as a starting point in understanding the growing process of NWs. An important growth theory on NWs was developed based on vapor–liquid–solid (VLS) with temperature ranging from 300°C to 1100°C.

The first step is the deposition of a catalyst required for the VLS process. This deposition normally occurs on a clean, planar topography and defect free surface. Materials that are often used are silicon wafers, glass or sapphire surfaces. The second step involves the melting of the catalyst that gradually forms a liquid droplet due to the surface tension. The diameter of the droplet, which is in the nanometer range, is the catalyst for growing the NW. With the continuous supply of gas precursor inside the chamber and the temperature of the chamber being kept constant, the material species (from the precursor) inside the chamber will keep falling onto the melted nano-droplets. As the parabolic surface of the nano-droplets have higher sticking coefficient than the bottom of the droplets, the concentrated incorporation of the material species (from gas precursor) will lead to hyper saturation of the material. The third step, alloying, will start to take place followed by nucleation of the material due to cooling and precipitation. Gradually, bit by bit, the NW is formed and grows upwards. Since the growing of NWs happens between the interfaces of liquid and solid and under continuous supply of material in vapor form, this mechanism is called vapor–liquid–solid or VLS. This mechanism can be summarized into three main steps as illustrated in Figure 1.2.

Heterostructure NWs can be grown with the same mechanism but with one or more layers of different materials being added horizontally or radially to the existing layer. This is done by alternating the precursors used and taking into account the difference between lattice constant of different materials [19, 20].

The diameter of the grown NWs corresponds directly to the diameter or the size of the melted catalyst nanodroplets. Since the nucleation and growing of NWs actually happen at the liquid–solid interface, the catalyst, which is in liquid form, will remain on top of the NWs at the end of the process. The NWs grown using this mechanism can reach a minimum diameter of less than 10 nm and a few microns or millimeters in length. Tuning and better control in synthesizing good quality of NWs via VLS mechanism however depends pretty much on constant growth rate, concentration of precursor, temperature, catalyst and time. Researchers have been applying various processes to grow the NW via VLS approach. Duan *et al.* [22] reported using a laser ablation technique. Laser pulses are used to vaporize a solid target containing the desired material and catalyst. Due to localized heating by the laser, liquid nanoclusters are formed and this nanoclusters will then define the size of the NW grown via VLS. Chemical vapor deposition (CVD) techniques such as metal organic CVD and low-pressure CVD were also used for the same purpose [23, 24]. By switching the chemical beam on and off repetitively over time, in a molecular beam epitaxy (MBE), SiNW growth has been reported by Schubert *et al.* [21]. Different combinations of semiconductor and metal growth via VLS mechanism are summarized by Fan *et al.* [4].

Selecting the correct catalyst material is crucial for growing NWs. This selection step is deduced from the binary phase diagram of the targeted material. By examining the phase diagram, the catalyst composition and temperature for growing NWs can be determined as reported by Duan *et al.* [3]. Epitaxial core–shell and core–multi-shell heterostructure NWs with different material and crystalline lattice were synthesized with the same technique as in single crystalline NWs [19, 25–28].

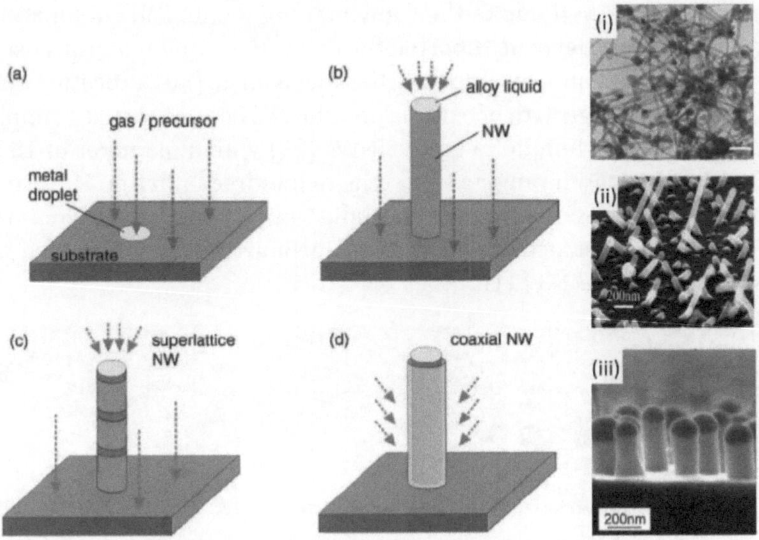

Figure 1.2 Vapor–liquid–solid growth mechanism: (a) Deposition of metal catalyst and formation of nano-droplets on a clean substrate. (b) Continuous alloying and nucleation resulting in NW growth at the pre-defined sites of nanodroplets formed previously. (c) Growing heterostructure NWs by alternating the gas precursors. (d) Coaxial grown NW by manipulating lattice matching. Examples of NWs grown via VLS with various technique of forming catalysts nanodroplets: (i) Laser ablation. (ii) Low-pressure chemical vapor deposition. (iii) Molecular beam epitaxy (reproduced with permission from refs. [3, 4, 21].

1.2.2 Non-Catalytic Growth: Vapor Solid (VS)

Alternatively, NWs can also be grown without the presence of a catalyst or the formation of nanodroplets to initiate the growth as explained in the VLS mechanism. It can be done via a thermal evaporation technique where the concentrated and condensed material is thermally evaporated at elevated temperature. Subsequent cooling or condensation of the vapor under correct and suitable operating windows (temperature, pressure and others) allows the growth of NWs. Since this mechanism of growing NWs doesn't involve the liquid phase, it is known as vapor–solid (VS) growth mechanism [4, 29, 30]. Si- and oxide-based NWs (such as Zn and Ga) can be grown using this method due to their low melting points and sublimation temperatures. Dai *et al.* reported the experiment and materials used to grow oxide nanostructures via this mechanism [30] indicating that the directional growth of oxide is possible. Yu *et al.* reported a simple evaporation technique to grow SiNWs [29] with a diameter of 15 ± 3 nm and a length ranging from tens to hundreds micron. Multistep and seeded growth of ZnO was studied by Liu and his co-workers. Complex nanostructures such as nanocombs were synthesized as shown in Figure 1.3 [31].

Figure 1.3 SEM images on complex nanostructures of ZnO synthesized. (a) Aka nano-turbines geometry. (b) Nanocombs. (c) Hexagonal nanorods (reproduced with permission from ref. [31]).

1.2.3 Template-Based Growth

In addition to VLS and VS gas phased growth of NWs, template-based growth through electrochemistry such as porous (AAO — anodic aluminum oxide), nanostructured polymers or mesoporous materials is another way for bottom-up growth of NWs [32, 33]. The first step is to make the porous or semi porous template. Then, the template is connected to a cathode electrode of a power supply. The electrode is then brought into contact with the material that is in the form of a solution. Similar to the electrochemical setup in a battery cell, the anode is placed in the deposition solution together or parallel to the cathode. Ye *et al.* described the growing of a cobalt NW via this process as explained in Figure 1.4 [32]. Yan *et al.* demonstrated that the AAO template is also suitable to be used as pattern transfer in wafer scale processing for sub-micron processing [34].

When connected to an electric source, ions inside the electrolyte or solution will attract (or repulse) depending on the electrodes polarity and subsequently a redox reaction happens. Cations will diffuse to the cathode and reduction happens, resulting in the growth of NWs inside the pores of the template. Once the pores are completely filled, free-standing NWs can be obtained by dissolution of the template membrane and the oxide layer. The NW diameter can reach 5–6 nm and the length up to 10 micron. The NW length can be tuned by the etching process and thus offering greater flexibility while processing. Due to its processing technique, NWs obtained from this technique are either amorphous or polycrystalline and mainly confined to metals [35]. Thus, NWs grown with this process may be limited in quality for some applications, especially in the field of optical devices.

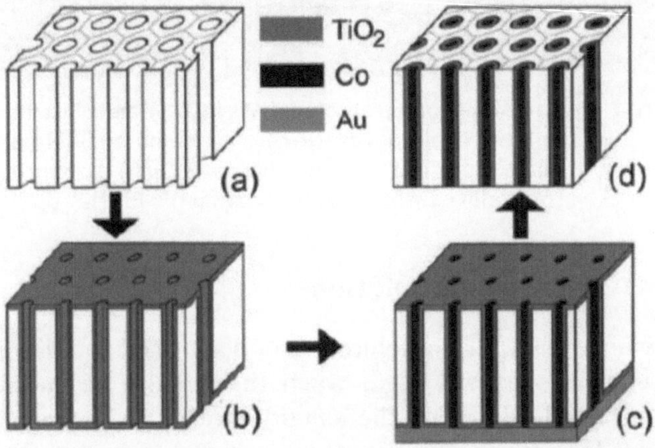

Figure 1.4 Schematic steps in growing NWs via template-based growth. (a) AAO porous template. (b) A layer of Titanium Oxide (TiO_2) is used to reduce the size of the template pores. (c) A gold layer is electroplated before the electrodeposition of Cobalt. (d) Cobalt NWs are formed inside the nanosize pores defined earlier (reproduced with permission from ref. [32]).

In 2010, Yarden *et al.* [36] demonstrated that complex geometries of NWs could be done using electrodeposition and Carbon Nanotubes (CNTs) patterns. First CNTs are grown into various patterns using self assembly. Then, using these patterns as the template for the deposition of NWs from a solution, large-scale processing, different shapes as well as specific shapes could be obtained. This is an interesting and not a rigid approach with huge flexibility in shaping the NWs. While parallel and straight NWs is common, having continuous and circular shapes such as coils, curves and serpentines are difficult and challenging to obtain. In general, any shape is possible and that is the reason, Yarden termed this method "drawing with nanotubes" as depicted in Figure 1.5.

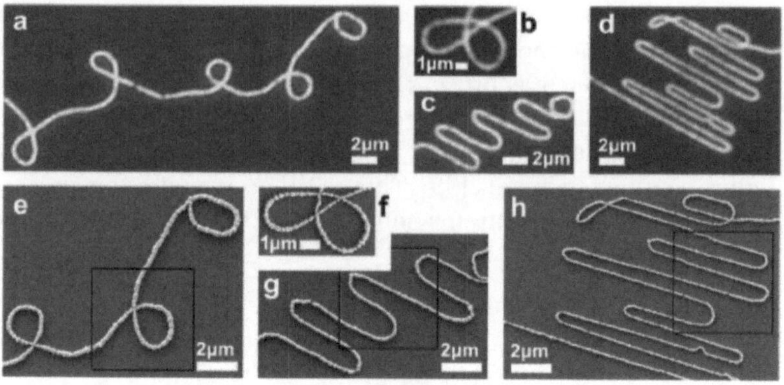

Figure 1.5 Complex geometries of Au NWs by CNT and electrodeposition: (a–d) optical images; (e–h) corresponding SEM images. (a,e) Coil shapes (b,f) pretzel shapes (c,d,g,h) serpentines shapes (reproduced with permission from ref. [36]).

1.2.4 Growing via Solution

Growing crystals from solution has been used for decades. The growing process will begin when the amount of the substance that is dissolved (called the solute) exceeds the solvent amount. This leads to over saturation of the solute dissolved in the solvent. Over time, due to precipitation and saturation of solvent, growth is observed. Using this method, materials from the II-IV groups are grown [4]. Holmes *et al.* [37] reported the growth of SiNWs from dodecanethiol-capped Au nanocrystals ranging between 20 and 30 Å in diameter using a solution-based synthesis. With a similar approach, Hanrath and Korgel [38] reported the successful growth of Ge NWs using alkanethiol protected gold nanoparticle/nanocrystals for seeding purposes. In general, solution-based synthesis results in smaller aspect ratios than those achieved by vapor phase synthesis. One advantage of the solution-based approach is the possibility of synthesizing NWs at low temperatures at large scales with low cost and on different substrates [39].

1.2.5 Orderly Growing: A Hybrid Approach

So far, the discussion on VLS, VS and solution-based growth focused on synthesizing NWs in the bulk size, not patterned and without any order. The template-based approach offers some degree of structure and order in the growth of NWs but lacks in the control of localized positioning. It generally produces irregular NWs, bad uniformity and is therefore difficult to use. The idea of controlling the position, arrangement and size of catalytic nuclei for growing NWs in an orderly manner has been reported by several groups [4, 16, 40, 41]. By taking advantage of the fabrication techniques where positioning and patterning can be easily done, NWs can be grown. This results in so-called hybrid approaches as top-down technologies are now combined with the bottom-up growth. Optical lithography [42], electron beam lithography [43, 44] and nanoimprint technology [45] have been employed to position and pattern the catalyst required for growing NWs. The main idea is to define smaller than 100 nm catalyst patterns (for example gold) for growing NWs via VLS mechanism. In this way, the grown NWs will have the same pattern and roughly the same diameter as the catalyst dots.

The main advantage of this technique is that the gold dots can be patterned on a large (square inch) scale with variable symmetry (square, hexagonal, and rectangular) and spacing. However, the NWs grown with this method are not straight and consist of cluster of NWs. A cluster of wires (up to 20) with diameters of 10–15 nm can grow from one defined Au site. The growing in clusters behavior is caused by the large size (50–200 nm) and thickness (15–20 nm) of the Au pads used.

Ng *et al.* reported on the nanopatterning and growth of ZnO NWs using e-beam lithography for a vertical surround gate field effect transistor (VSG-FET). Martensson *et al.* [43] also reported the same activity but with InP NWs grown with MOVPE from gold catalysts defined by e-beam lithography. The advantage of using e-beam lithography is the good control for the separation of NW growth sites, while the major drawbacks are the high cost involved and the time consumed for the e-beam writing. Figure 1.6 shows the utilization of a hybrid approach where catalysts were pre-patterned via e-beam lithography before the growing step.

Several unique but interesting approaches have been implemented. Ohlsson *et al.* [46] succeeded in manipulating single gold nanodots on the substrate surface for synthesis of selected size GaAs epitaxial NWs using nano-manipulation of aerosol particles by means of atomic force microscopy. Lieber and coworkers [47] manipulated the suspensions of gold nanodots in colloids of different concentration for nanopatterning before growth. The advantage of using a gold nanodot suspension is the size control, without the need of using lithography and yet with dot diameter around 50 nm. Recently, Yang *et al.* [48] reported epitaxial SiNW growth on Si substrate using gold colloids immobilized by coating the Si surface with a thin layer of polyelectrolyte, which acts as electrostatic attraction to Au colloids. Nanoimprint has becoming more prominent nowadays. For example, McAlpine [40] reported on using nanoimprint lithography for precisely positioning the catalyst for NW growth via imprint and etching using a Si/SiO_2 stamp.

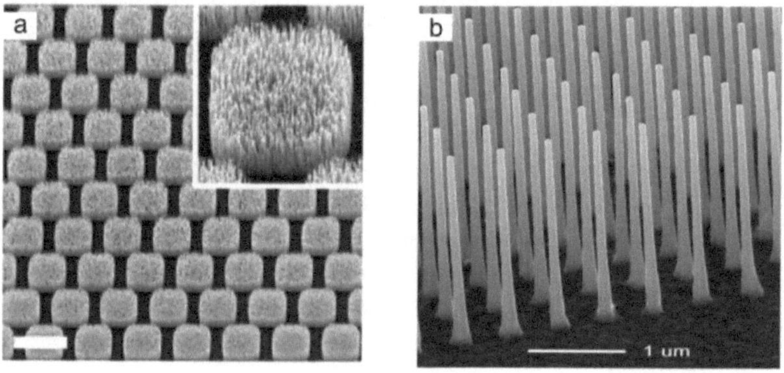

Figure 1.6 NWs grown on patterned catalyst using electron beam lithography before the growing process. (a) Vertical ZnO NW grown on pre-patterned SiC substrate (reproduced with permission from ref. [44]. (b) InP NW grown with gold catalysts defined by e-beam lithography (reproduced with permission from ref. [43]).

In 2005, Fuhrmann [16] applied the concept of nanosphere or natural lithography in growing SiNWs. In this technique, a monolayer or bilayers of spheres were first deposited on the Si substrate. This monolayer and bilayers will act as the soft mask for the patterning of gold, which is the catalyst for SiNW growth. SiNWs will then be grown using molecular beam epitaxy. Figure 1.7 shows the steps of SiNW synthesis using nanosphere lithography. An unique advantage of this mask transfer technique is the transfer process is relatively easy, not limited to one kind of substrate and the gap between spheres provides good reduction in size down to 30 nm with reasonably good separation between the deposited gold nanodots.

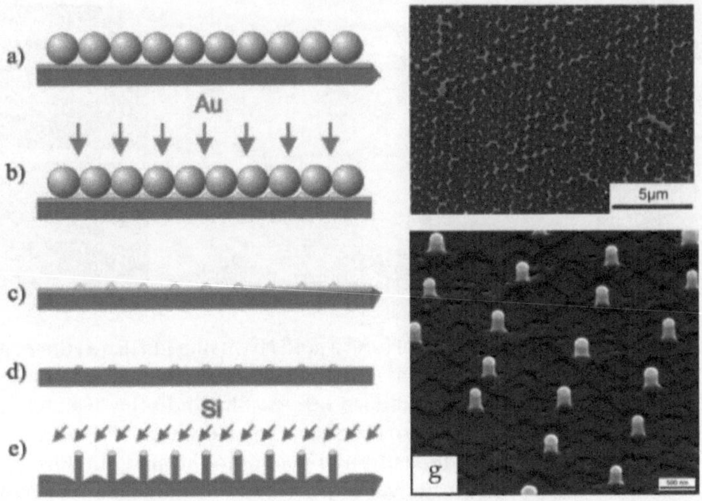

Figure 1.7 NW fabrication via Nanosphere lithography: (a) Deposition of a single layer of polystyrene beads forming a mask layer (b) Deposition of gold catalyst by thermal evaporation, (c) Removal of the masking layer (d) Thermal annealing and cleaning step to remove the oxide layer, and (e) SiNW growth by MBE. (f) Corresponding SEM image of step (c) (g) SEM image on NW arrays grown via Nanosphere lithography and MBE. (reproduced with permission from ref. [16]).

Another hybrid approach is based on electrodeposition of metals on a pre-defined undercuts. This method has been used to make metal NW from different materials such as platinum [49, 50], gold [51, 52], palladium [50], silver [53], bismuth [50] and others. Xiang *et al.* [50] described an approach called lithographically patterned NW electrodeposition (LPNE). LPNE involves photolithography of a temporary, sacrificial film on the surface to create a small gap composed before electrodeposition of a metal film is carried out. The process flow and results of LPNE are represented in Figure 1.8. This method produces a generic rectangular cross section NW with a flat top with a width ranging from 11–30 nm to few hundred nanometers depending on the type of metal and electrodeposition parameters. However, the deposited NW is not completely straight and lots of edges and small curves could be observed as shown in Figure 1.8e,f. The height of the NW is determined by the vertical length of the undercut or the thickness of the deposited sacrificial film (Figure 1.8b).

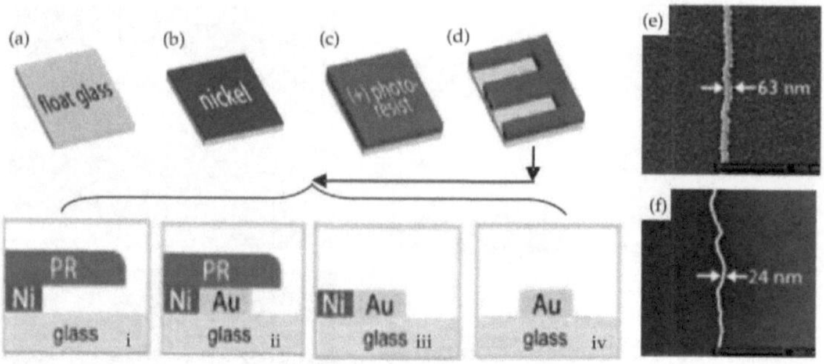

Figure 1.8 An example of patterning a gold NW using LPNE. (a) Clean glass substrate. (b) Thermal deposition of nickel. (c) Spin coating of photoresist. (d) Standard lithography and selective etching of the exposed nickel to produce a small undercut as shown in the (i). (ii) Electrodeposition of gold onto the small undercut; (iii) and (iv) After removal of photoresist and nickel, respectively. (e) and (f) corresponding NWs produced via LPNE (reproduced with permission from ref. [50])

1.3 Top-Down Approaches

1.3.1 Beams Lithography

Instead of making complex structure from atomic and molecular level, top-down processing is often associated with adding layers of different material on each other via deposition and subsequently removing the unpatterned region via etching in a plasma or in a solution. In the nano world, for tiny and small dimension structures, the easiest way is to use electron beam (E-beam) lithography or focus ion beam (FIB) lithography. The size of electrons and ions are very small. By manipulating the beam of electrons and ions, dimension as small as few nanometers can be patterned on a layer of resist. This pattern is then transferred into the substrate material by etching. In most of the cases, the substrate material is silicon.

Several research groups have been using e-beam lithography. Among them, Li *et al.* [10] first demonstrated precisely controlled fabrication of SiNW devices using E-beam lithography with sub tens nanometer SiNWs using silicon on insulator (SOI) wafers. Trivedi *et al.* [54] in 2011 reported using e-beam lithography and an oxidation process to reduce the size of the initial structure (sub-microns) to sub-5 nm NWs. Tettamanzi *et al.* [55] demonstrated an alternative way by using FIB lithography to fabricate Niobium (Nb) superconducting NWs. However, this type of patterning (E-beam and FIB) is very slow, time consuming and expensive compared with conventional optical lithography. It is believed that conventional optical lithography in combination with a size reduction strategy may provide a simple and economic solution to realize SiNWs for commercial devices [13, 50, 56, 57].

In a recent publication, a new maskless method of patterning using beam lithography emerged. This method is called laser interference patterning (LIP) [58, 59]. It is known that a laser beam can be split into coherent beams with lenses and a beam splitter. By exploiting the standing wave pattern produced by the interference between these beams, parallel and large-scale patterning is possible [60]. The interference period is controlled by the wavelength and the angle of the beams [58]. Figure 1.9 shows the schematic representation of the experimental steps of LIP. After double patterning, nanodots are formed and subsequent dry etching will form vertical NWs. Although the patterns are limited to squares and hexagonal patterns, this method could potentially scrub the main disadvantages faced by using E-beam or FIB patterning. Nevertheless, direct LIP is capable of achieving nanometer wide NWs but so far, only few micrometer long NWs are produced [59].

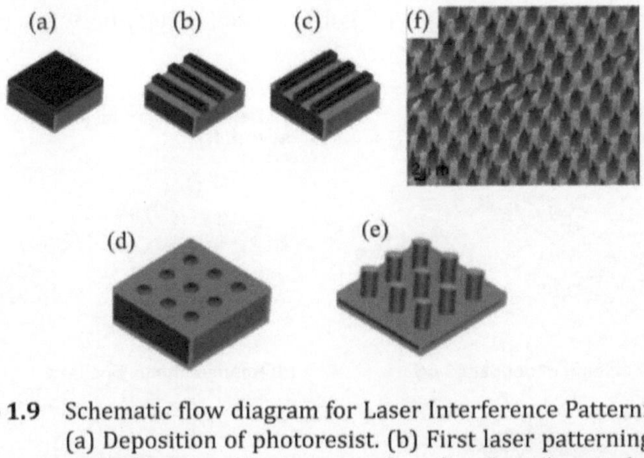

Figure 1.9 Schematic flow diagram for Laser Interference Patterning-LIP. (a) Deposition of photoresist. (b) First laser patterning of the resist (c) Second laser patterning after the substrate has been rotated 90°. (d) Formation of nanodots (mask). (e) Vertical NWs after subsequent etching into the substrate (f) Vertical acetate film NWs patterned via LIP (reproduced with permission from ref. [58, 59]).

1.3.2 Spacer/Edge Transfer Lithography

Although it has been elegantly demonstrated that well-defined SiNW devices can be fabricated in the previously discussed top-down approach, some fundamental top-down processing targets, such as low-cost, high-yield and fast fabrication processes, remain unachieved. To use SiNWs in commercial devices [13, 50, 56, 57], Choi *et al.* [57] introduced a technique named spacer or edge transfer lithography. This method is purely based on microfabrication technique and does not require any nanolithography tool.

The beauty of this technique is based on the fact that deposited material on non-planar surfaces covers the step edge as well as the top of step as explained in Figure 1.10. In contrast, plasma etching is an anisotropic technique that removes materials preferentially in the direction perpendicular to the surface. Due to the fact that the thickness of the deposited film can be controlled down to 10 nm or less with high precision and very conformal coating, this technique is capable of generating nano-scale pattern that would not be possible using optical lithography [57, 61]. As illustrated in Figure 1.10, the first step for spacer fabrication is the deposition and patterning of a sacrificial or dummy structure. The width of the dummy should be bigger than the desired width. After patterning the sacrificial layer, a different and selective removable material compared with the dummy is deposited (Figure 1.10b). The selectivity between the two materials should be high.

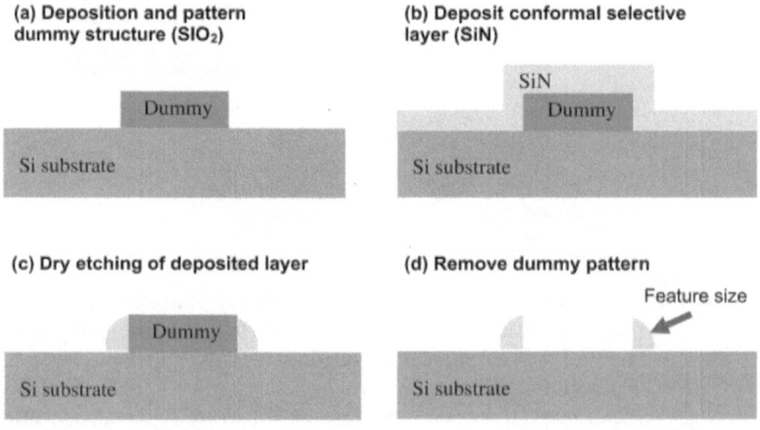

Figure 1.10 Conceptual drawing on spacer/edge transfer patterning of NW (adapted from ref. [57]).

Dry etching follows after the deposition step. While the deposited material that is perpendicular to the surface will be etched, deposited material on the sidewalls will not be etched (Figure 1.10c) in the dry etching step. Removal of the sacrificial layer/dummy is usually done by wet etching. Due to the high selectivity between the deposited and sacrificial/dummy material, the dummy/sacrificial layer will be rapidly etched away. The left-over of the deposited layer on the edge of the dummy structure (Figure 1.10d) will act as the masking material for patterning the layer underneath. This process is known as spacer formation. The spacers define the width of the desired NWs [61]. Therefore, the thickness of the deposited layer will determine the feature size (width) of the intended structure. Continue etching downward into the layer underneath will form the desired small and sub-10 nm NWs.

With this technique, dense, conformal, less variation in dimension, large scale, low cost, fast processing could be achieved. However, the pitch in processing NW via this technique is limited to the average pitch of the spacer/width, usually half of the dummy/sacrificial feature [62]. This pitch can be increased via double or quadruple spacer processes [63]. This spacer technique offers a better uniformity while patterning smaller feature size compared with E-beam lithography [61].

1.3.3 Nanoimprint Technology

Nanoimprint lithography (NIL) has been increasingly important for high-throughput patterning of nanostructures [45, 64, 65]. The idea behind is to emulate the molding industry where using a single mold, casting and cooling, an identical structure can be reproduced on a very large scale. There are two major steps in nanoimprint lithography: stamping and printing. The stamping step refers to the pre-prepared and well-defined patterns inside a mold being brought in contact with a resist layer on the substrate surface. Printing involves transferring the patterns in the mold onto the substrate. These steps are repeated periodically to create a very high yield and throughput patterning. Available methods for NIL are step and flash imprint lithography (S-FIL) and thermal NIL. The S-FIL method requires the use of a low viscosity photo-curable monomer to be coated on the surface of a substrate. Then, a transparent stamp, which can be made from fused silica, is brought into contact with the monomer. After UV exposure, patterned polymers by the mold (due to the force applied while stamping) will be cured and hardened for further process. In thermal NIL, the stamp used should withstand high temperature during imprint as heat will harden the patterns formed by the mold. Thermal NIL offers more flexibility in choosing material for the mold as any hard material could be used since transparency is not an issue in this case. Electron beam lithography is the most standard way to fabricate the stamp with high resolution. However, different methods to fabricate molds for NIL such as edge lithography [45], shadow deposition patterning [64] and mold-to-mold cross imprint (MTMCI) [65] have been reported recently.

1.3.4 Oxidation

A silicon surface will oxidize into silicon dioxide when exposed to oxygen. However, under room temperature the oxidation process is slow and at certain level the growth of the oxide will saturate [66]. Thermal oxidation provides a faster route to grow silicon oxide and thus, this technique is very suitable for IC compatible processing, allowing massive and large-scale processing of devices. By heating up the Si substrate in the range of 800–1200°C and feeding the chamber with either pure oxygen or water vapor flow, the oxide growing process is initiated. If the silicon surface is coated with a silicon nitride layer, no oxidation will take place. This allows local oxidation of the silicon, also known as local oxidation or LOCOS [67]. This selective oxidation was used by Pennilli *et al.* [68] to reduce laterally the size of SiNWs as shown in Figure 1.11.

One of the minor disadvantages of using this technique is the formation of a so-called bird's beak structure at the corner interface between silicon and the masking materials (silicon nitride) during the oxidation process [70]. The formation of this beak will cause the structure to be widened particularly on the edge. This phenomena is explored by Chen *et al.* [69] whereby it was reported that only a slight deviation occurred in the structure (only 6.4° difference in the angle measured between major crystalline planes). This slight difference is caused by the non-uniform growth at the interfaces. The bird's beak phenomenon is illustrated in Figure 1.11f.

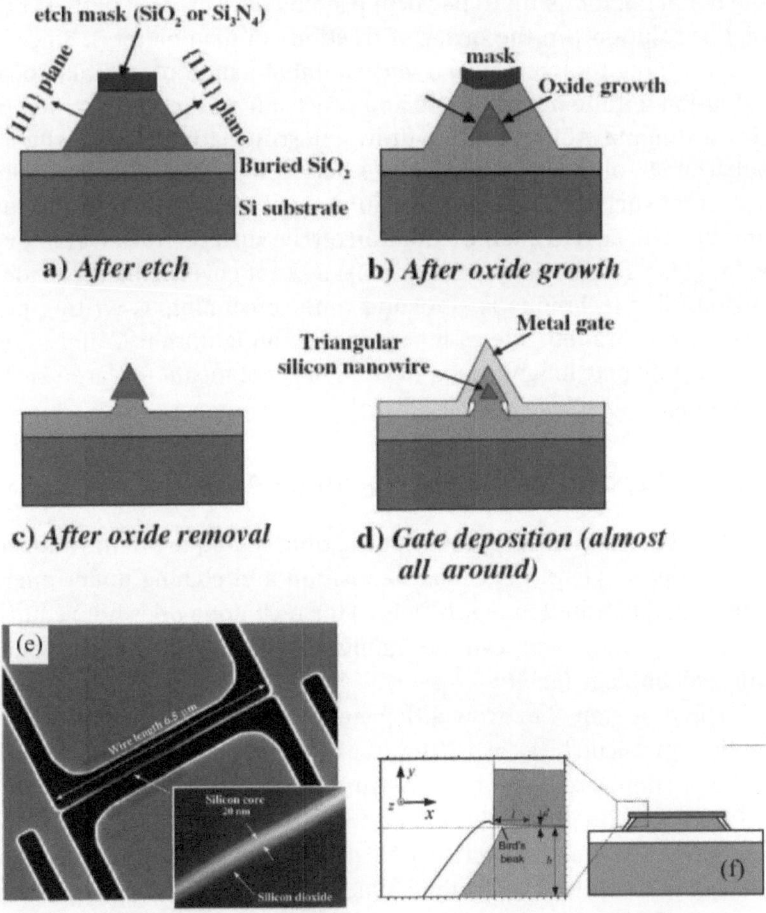

Figure 1.11 Fabrication of SiNWs via oxidation reduction scheme. (a) Definition of the initial NW structure by anisotropic wet etching. (b) Thermal oxidation. (c) Selective removal of the oxide layer, revealing the triangular shape SiNW (shape based on initial structure) (d) Formation of metal gate for FET application. (e) Oxidation fabricated SiNW with inset showing the 20 nm SiNW. (f) Sketch showing the formation of the bird's beak structure after the oxidation (reproduced with permission from refs. [68, 69]).

Recently, Martinez *et al.* [71] reported on direct oxidation by writing on Si substrate via atomic force microscope (AFM) lithography or nanotip lithography. AFM is a scanning probing tool that uses the deflection of its tip to perform probing over an area with a very high resolution (on the order of fractions of nanometers). Since it is a scanning tool, it is also a very suitable nanotool to make local oxidation with nanometer size and precision on the Si substrate. In this technique, AFM is used to draw a narrow pattern on the silicon substrate by applying voltage pulses between the AFM probe and the substrate surface. These voltage pulses will form a water meniscus and this will lead to anodic oxidation on the surface. Next, wet or dry etching is used to remove the unmasked (not covered by the oxide) region. This technique is slow and time consuming as writing per scan size is limited, due to its hardware configuration. Wafer scale writing will take longer times since AFM is not meant for large-scale scanning.

1.3.5 Deposition and Etching under Angle

Another low-cost and top-down fabrication technique was introduced by Hien *et al.* [13]. It is called deposition and etching under angle (DEA) as illustrated in Figure 1.12. This technique does not require nanolithography and can be implemented in any conventional microtechnology facility.

The first step is to grow a thermal oxide layer by wet-oxidation on the top silicon layer of a SOI wafer. Then, low stress silicon nitride (SiN) is deposited via low-pressure chemical vapor deposition (LPCVD) on the grown SiO_2. A window in the SiN layer is opened using reactive ion etching (RIE). Next, the exposed SiO_2 layer is wet-etched with diluted hydrofluoric acid (5%) to form an undercut region between SiO_2 and SiN, which is the cavity. The etch rate of oxide in hydrofluoric acid is known and controllable [72]. A thin layer of chromium (Cr) is then electron-beam evaporated at 45 degrees with respect to the substrate. Excessive evaporated metal outside the cavity is trimmed using ion beam milling at –45 degree. The remaining SiO_2 and SiN layers are then removed by etching and the earlier deposited metal inside the cavity is not removed or altered. The SiNW is formed when RIE etching is carried out with the pre-defined metal hard mask.

Using this DEA technique, the lateral width of the fabricated SiNWs is around 33 nm up to several hundreds nanometers, with a minimum length of several micrometers as in Figure 1.12j. The

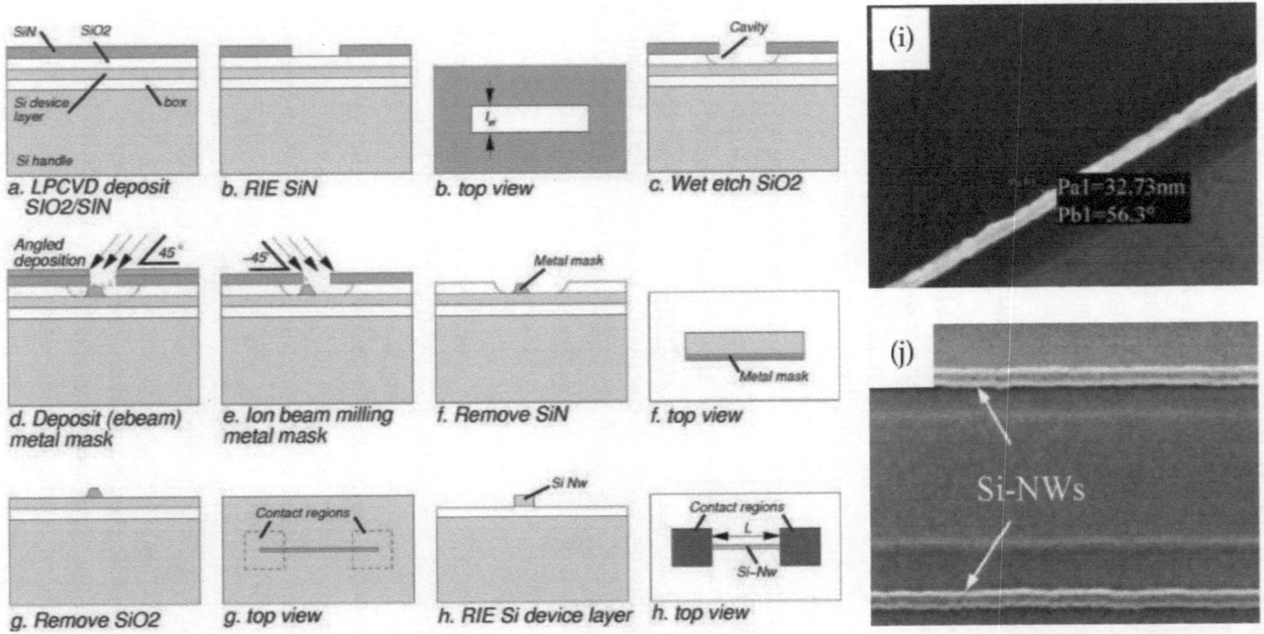

Figure 1.12 Single mask SiNWs deposition and etching under angle fabrication process (steps a–h). (i) Evaporated thin metal film (width 33 nm) after stripping of oxide and nitride layers (j) Fabricated parallel SiNWs on SOI wafer (reproduced with permission from ref. [13]).

width of deposited metal mask will determine the width of the SiNW after etching. The quality of the fabricated SiNWs could be improved by optimizing the ion beam milling angle and deposition of the metal mask layer such that the metal layer is located mostly inside the cavity. The larger milling angle will produce a well-defined edge and reduce surface roughness, (surface roughness is very important in chemical functionalization and sensing application).

1.3.6 Anisotropic Etched Microfabrication SiNWs

Microelectromechanical systems (MEMs) technology is reaching a good level of maturity. It provides a good platform to co-fabricate nanodevices together with CMOS or IC compatible processing. This leads to potential large-scale, low-cost, orientation dependant, smaller but high-yield production of nanodevices such as SiNW field effect transistor (FET) devices. Surface and bulk micromachining play an important role in MEMS and nanofabrication. Surface micromachining builds microstructures in different layers on top of the substrate via selective etching and deposition. On the other hand, bulk micromachining involves selective etching inside a substrate to build microstructures. In the fabrication of SiNWs, both processes are useful in creating many small scale devices. It is not an easy task but involves creative and innovative ways of processing. One of the key features of bulk micromachining is the anisotropic wet etching of single crystalline silicon [73]. It provides an alternative approach in fabricating NWs particularly on silicon or SOI wafers.

Typically, wet (solution) etching is an isotropic event. When the etchant (solution) is being brought in contact with the layer to be etched, a chemical reaction occurs and produces soluble by-products, which will be transported away from the substrate. Isotropic etching refers to non-directional etching and has constant and uniform etching rate in all direction. Nevertheless, this generic concept does not always apply to crystalline materials. In a crystalline material such as single crystalline Si, the etch rate is typically lower in the densely packed lattice compared with loosely packed lattice. Si has a diamond like lattice structure as described using the miller indices [74]. The density of atoms on a Si wafer differs according to the planes with {111} being the densest followed by {100}. Meanwhile,

{110} planes are the loosely packed plane. Thus, the etching rate in a wet etchant (solution) differs depending on crystalline orientation with {110} as a faster etching plane followed by {100}, while {111} is the slowest etching plane.

By taking the advantage of the different etching rates between different crystalline planes, orientation dependent etching and controllable etching rate, nanoscale devices could be fabricated without the need of expensive and time consuming advance lithography such as E-beam and FIB lithography.

Up to date, most of the top-down silicon NWs fabricated via conventional top-down microfabrication techniques offer one or two crystalline surfaces/planes [12, 69]. Stern *et al.* [12] proposed a process for trapezoidal shaped silicon NWs with {111} and {100} surface planes fabricated with E-beam lithography and anisotropic etching. Meanwhile, Pennelli *et al.* [68] fabricated SiNW via anisotropic etching resulting in two {111} planes followed by lateral oxidation to reduce the dimension of the SiNW. In this section, only SiNW fabrication processes that are based on microfabrication or surface micromachining will be discussed.

Recently, Chen *et al.* [69] demonstrated the fabrication of sub-30 nm single crystalline {111} planes SiNWs using the technology illustrated in Figure 1.13. The starting point of this process is the deposition of the LPCVD silicon nitride layer on top of a SOI wafer. SOI wafer is important because the buried oxide layer will act as a stop etch layer and thus it is easier to control the structure as well as the dimension of the etched structure. A first lithography step is done by aligning the mask along the [110] crystalline plane. The wafer is etched using tetra methyl ammonium hydroxide (TMAH). The etching in [111] plane is significantly slower compared with other planes. Virtually, the [111 plane would be considered as the planes where the etching stops. Then, oxidation is done to create an oxide layer on the newly etched surfaces while retaining the nitride layer. A second lithography step is done to open a smaller window on top of the first patterned structure and thus exposing the silicon layer underneath the nitride layer. A second anisotropic etching step is carried out and the etching will stop on the [111] plane. After removal of the nitride and oxide layer, two triangular shape SiNWs with two [111] planes side by side existing in parallel are clearly visible as shown in Figure 1.13e.

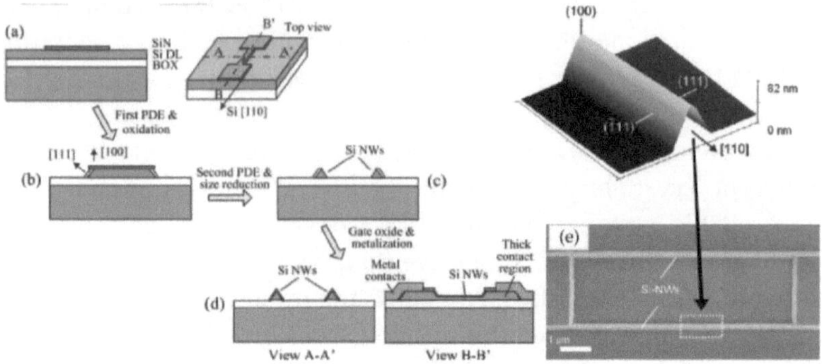

Figure 1.13 Crystalline plane dependent etching process. (a) Mask being aligned to [110] plane. (b) First anisotropic etching step and stop on {111} planes followed by oxidation. (c) A window is opened and a second anisotropic etching is carried out. (d) Cross section and elongated view of two {111} planes after stripping off the mask. (e) SEM image on parallel SiNWs (reproduced with permission from ref. [69]).

In the Delft Institute of Microsystems and Nanoelectronics (DIMES) laboratory at the Technical University of Delft, The Netherlands, we developed an alternative yet unique process to fabricate SiNWs with three {100} surfaces using solely anisotropic etching and conventional microfabrication facilities [75]. This process offers an upgrade in the number of surfaces available, i.e. three compared with the two surfaces available via other technique discussed. Offering the extra number of surfaces is not the only advantage of this technique. It also offers the precise control of the width (sub-100 nm), results in smooth and straight sidewalls, and requires a room temperature etching and a single mask process. The main essential point in this technique is controlling the width of SiNWs via etching time. With a very slow etch rate, good control on the final dimension is possible.

The SiNWs are defined using conventional lithography and anisotropic wet etching on a (100) SOI wafer. The fabrication process is schematically shown in Figure 1.14. A 100 nm plasma enhanced chemical vapor deposition (PECVD) oxide and 600 nm PECVD nitride are first deposited and dry etched resulting in a thin rectangular slab with the width of 1.2 μm and a length of 5 μm. The pattern is aligned parallel to the <100> plane, as indicated in Figure 1.14a. Then, the wafer is immersed in etchant (AZ400K developer) at room temperature for 75 minutes. The developer etched silicon both vertically down into the wafer and also laterally underneath the masking oxide. The vertical etch will stop at the buried oxide (BOX) layer, while the lateral under-etch will continue inwards, further decreasing the dimension of the wires. Due to the mask being oriented 45° to <110> direction, the lateral etch front will encounter {100} planes on the sidewall of the structure.

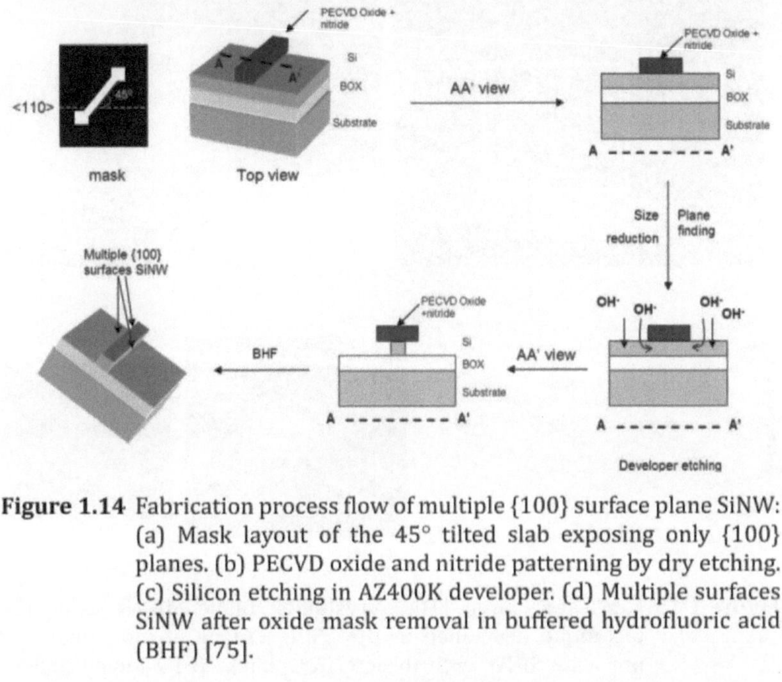

Figure 1.14 Fabrication process flow of multiple {100} surface plane SiNW: (a) Mask layout of the 45° tilted slab exposing only {100} planes. (b) PECVD oxide and nitride patterning by dry etching. (c) Silicon etching in AZ400K developer. (d) Multiple surfaces SiNW after oxide mask removal in buffered hydrofluoric acid (BHF) [75].

As for silicon etching, we use a different etching solution compared with others. A standard and widely available AZ400K developer with the composition of 15% Potassium Borate (K_2BO_3) in water is preferred to be used compared with TMAH or KOH solutions, which are usually known. This is due to the slower, but constant etching rate (Figure 1.15c) that makes it easier to control the final width of the NWs in a reproducible way.

The starting material is a (100) SOI wafer, which has a 340 nm thick single crystalline silicon layer and a 400 nm buried oxide (BOX) layer. Since the top surface of the starting wafer has (100) crystalline plane, the etched two sidewalls are also {100} resulting in a SiNW with three {100} surface planes. As shown in Figure 1.15b, these etched sidewalls are straight and smooth. With this process, we managed to decrease the width of the wires from 1.2 μm to only 100 nm (Figure 1.15d) while keeping very smooth surfaces and well-dimensioned NWs.

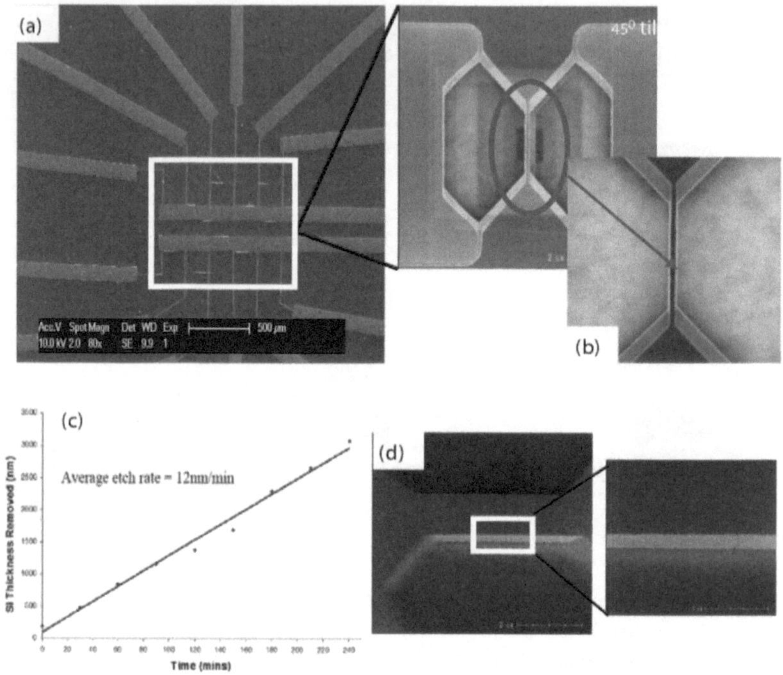

Figure 1.15 Fabricated three {100} crystalline plane SiNWs using the technique developed in DIMES. (a) Array of individual 100 nm wide SiNW with three {100} planes. (b) Zoomed images on an individual 100 nm SiNW on the right. (c) Crystalline silicon etch rate in AZ400K developer (d) SEM images showing {100} straight and smooth plane sidewalls after etching in the developer solution [75].

1.3.7 SiNW Field Effect Transistor

The starting point in downsizing the electronics components from micro- to nanoscale ignited the flurry in biosensing researches and applications as well [76, 77]. Current development on 1-D nanostructure clearly has its own advantages, and yet, the question is: why do we need nanoscale for biosensor development? The answer lies in the sensitivity of the sensor. High surface-to-volume ratio, tiny dimension and highly sensitive yet selective modified surface are certainly ideal platform to work with. Indeed, the miniaturizing of device dimensions not only improves the density packing of the device but also decreases the power consumption. Yet, a major gain in sensitivity to surface absorption is achieved dramatically compared with bulk structures [14, 18].

Semiconductor FET NWs for biosensing have gained quite some attention. Sensing here refers to a platform that is able to detect the signal or quantity of interest. Often, this platform involves surface chemistry for selective detection. Meanwhile, the transducer is a platform that enables the sensor to relay the detected signal to the outside world into readable or recordable form such as an electrical signal. Thus, the FET in the sensing setup is the transducer, meanwhile the SiNW together with the help of surface chemistry is the sensing platform [5].

Nair *et al.* and Alireza Kargar *et al.* reported mathematical formulation and simulation of SiNW FET sensors focusing on different geometry and its sensitivity [78, 79]. Similarly, Niklas Elfstrom *et al.* investigated the connection between sensitivity and width and thickness of the NW FET devices [14, 80]. Sensitivity is often defined as conduction ratios between initial and subsequent conduction as the NWs surface charges are altered through DNA, viruses and other molecules [5, 81, 82]. While the principle of detection lies in the changes of the conductance inside the NWs, the sensitivity differs between metallic NWs and semiconductor NWs. Metallic NWs lack in sensitivity compared with the semiconductor FET NWs [13]. In a metal oxide field effect transistor (MOSFET), the conductance inside the channel is the highest where changes in the drain current is the highest or the area around the threshold voltage [83, 84], and consequently where the sensitivity is the highest as the majority carries inside the channel gains further momentum (Figure 1.16b). Therefore, any changes on the surface of NWs will be detected. However, it is not clear as for now which detection mode is more sensitive or better [13].

Dopants such as boron or phosphorous are used to introduce impurities inside the Si crystal lattice to modulate its electrical properties. Nair *et al.* [79] studied the doping effects on the sensitivity. Lower doping concentrations (10^{16}–10^{17}) will increase the sensitivity of SiNW FET devices and higher doping ($>10^{19}$) will result in semiconductor NWs acting as metallic NWs [13]. Niklas *et al.* reported 2-D simulation on the sensitivity of Silicon NW FET [14, 80]. It is observed that for smaller NWs (width $w \leq 140$ nm) there is a quenching of the channel due to surface charges at low drain-source voltage bias, V_{DS}. As the width becomes too small, the surface charges are blocking the channel and this phenomenon increases with decreasing of the NWs width.

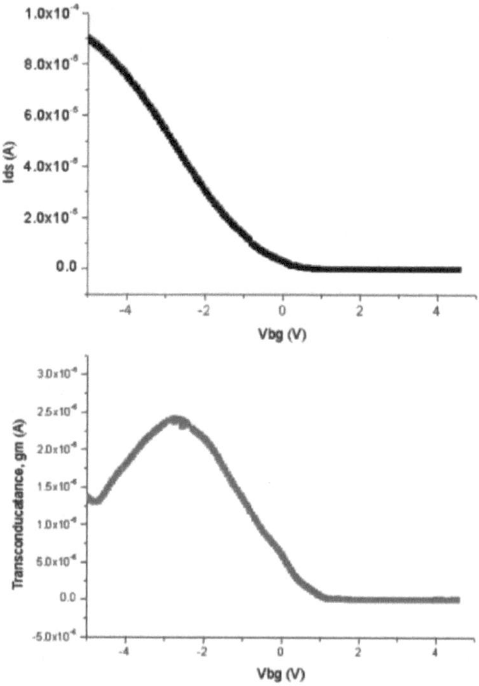

Figure 1.16 (a) The characteristic curve (drain–source current, I_{ds} versus backgate voltage, V_{bg}) of a SiNW FET device with V_{ds} of 2.5 V. (b) Corresponding transconductance, g_m curve for the same device [75].

For the NWs ranging from 140 nm < w < 160 nm, a similar characteristic to a MOS transistor is detected. And for larger width, w > 160 nm, higher gate voltage and bias are needed. It is also important to note that the threshold voltage is dependence on the NWs width; the smaller the NWs width is, the larger the threshold voltage. Stern *et al.* noted that the smaller NWs width with dimensions varying from 25 nm to 80 nm, exhibits constant transport properties throughout the NWs [12]. Table 1.1 summarizes the doping level on SiNWs that is available in literature for biosensing purposes.

Table 1.1 Comparison of doping type and concentrations according to the fabrication techniques used

First Author	Fabrication techniques	SiNW dopant concentration (cm^{-3})	References
Fernando Patolsky	CVD-VLS	P-Type, Boron, 1:4000 of B/Si ratio	[5, 11]
Yi Cui	Laser ablation-VLS	P-Type & N-Type, Boron and Phosphorous doped *(NA)	[17, 47, 85, 86]
G.B. Kang	Stepper + size reduction	P-Type, Boron, 10^{18}	[87]
X.T. Vu	Nanoimprint	P-Type, Boron, 10^{14}	[88]
Zhiqiang Gao	Oxidation	P-Type, Boron, $\sim 10^{16}$	[56]
J. Martinez	AFM oxidation	N-Type, Phosphorus, 10^{18}	[71]
Niklas Elfstrom	E-Beam Lithography	P-Type, Boron, 10^{14}	[80]
Eric Stern	E-Beam Lithography	P-Type, Boron and Arsenic doped, $\sim 10^{15}$	[12, 89]
G. Pennelli	Lateral Oxidation	Spin on dopant, P-Type and N-Type *(NA)	[68]
H.D. Tong	DEA	P-Type, Boron, 10^{16-18}	[13]
Songyue Chen	Anisotropic {111}	P-Type, Boron, 10^{17}	[69]
T.S.Y. Moh	Anisotropic {100}	P-Type, Boron, 10^{16-17}	[75]

* NA = Exact value not available.

1.4 Top-Down Versus Bottom-Up Approaches

Size does matter. By using bottom-up synthesis, very small dimension of NWs, for example as small as a few nanometers, is relatively easy, less rigid in processing control and feasible to achieve. In terms of number of NWs, definitely the bottom-up approach has an advantage over the top-down approach, and indeed, for the application than requires bundles of NWs or canopies of NWs, it is more appropriate to use a bottom-up approach. Moreover, very sophisticated and advanced equipment is not required. .However, NWs grown using bottom-up synthesis have one feature in common. Due to the growing mechanism NWs are reaching outward, mostly vertically, and exist in bundles on the substrate. While protruding NWs are suitable for open sensing, for example, gas sensing, and do not required complex device integration [90], doping can be a great challenge in term of better control of dopants concentration and dopants distribution uniformity [5]. Having the NWs standing vertically restricts the integration of NWs to form a functional device as most of the electronics components are assembled in a horizontal scheme. Sticking out or protruding structure will limit the device packaging possibilities and takes up unnecessary large areas. Therefore, having a single NW and integrate it into a functional device such as a biosensor poses great challenges.

In the literature, it is reported that assembling NWs grown via bottom-up approach often requires a "pick and place" procedure [15]. From the bundles of NWs grown, often picking and selecting one functional and good shape NW is tedious and time consuming. Since most of the grippers/tweezers are either in micro or millimeter size, assembling, moving and positioning NW is an extremely challenging operation. In addition, NWs can't be seen with naked eyes as well as in an optical microscope and thus often rely on SEM or TEM for handling and inspection. Without a good way to assemble the ready grown NWs, manipulating and designing suitable experiment setup/tools is nearly impossible.

Duan *et al.* [91] first proposed a solution to this problem by using directed electrical field to assemble the NWs. An electric field is applied to position and align the NWs by attraction and repulsion forces due to difference in polarities. This technique however has one big problem; the fringing effect in the electric field, especially at the submicron scale, hinders the flexibility in the assembling process. This limits the range of positioning of the NWs. Fluid flow assembling, which is taking the idea of flowing the logged tree trunks down the river to the processing plant downstream, allows more flexibility for assembling NWs as reported by Huang *et al.* [15]. Grown NWs are floated inside a small fluidic chamber and by directing the flow, alignment and positioning NWs at desired location is achieved.

Assembling via electric field or fluid flow is easier to accomplish with bundles of wires, while it still poses great difficulties to get a single NW in place. The hybrid approaches discussed in Section 2.5 could potentially individually address the issue of localized growth of NWs but the growing process of NWs often do not yield the best uniformity, shape, dimension and orientation of the NWs.

The challenges and difficulties in assembling synthesized NWs via bottom-up approaches require an alternative way of processing NWs and subsequently integrate them into functional devices with ease, high yield, low cost and efficient processing time. Top-down approaches provide a more robust, mature and convincing approach as it utilizes the more advanced and mature technology available in the semiconductor field, such as lithography patterning and etching. By riding on the advancement wave in IC compatible processing technology, this approach has several advantages to offer. Table 1.2 compares all bottom-up and top-down fabrication features of NWs discussed in this review.

1.5 Conclusion and Future Challenges

There are two broad approaches to make NW-based sensor. The question, of which one is better or most suitable to use, does not get a straightforward answer. In general, a combination of elements in both approaches might give the best results. Mainly, the most suitable method will be related or dependent on the application these NWs will be used for. For the bundles like and vertical orientated applications, a bottom-up approach is better as it offers very small dimension in width and a large flexibility in shapes. As for ordered, uniform, high density of NWs, better control on dopants concentration (FET devices), easy integration and packaging, the top-down approach is more suitable. Even though e-beam and AFM lithography provides the easiest path to achieve small dimensions, the cost is high, the throughput yield is low and the process is time consuming. Conventional microfabrication together with anisotropic etching offers a promising alternative for fabricating NWs with high yields, easy integration, large scales and better controls of process parameters such as doping levels. Although the width of NWs fabricated via this technique is not as narrow as NWs fabricated via e-beam lithography, they are sufficient for the purpose of biosensing as discussed above. More research and optimization of current techniques are needed to achieve ultra small dimension and better performance of top-down fabricated NWs, while paying attention to reliability, stability and repeatability.

Table 1.2 Comparisons between bottom-up and top-down approaches toward NW fabrication

	Bottom-up approaches				
Dimensions	**VLS**	**VS**	**Template growth**	**Growing via solution**	**Hybrid growth**
Width/ diameter (nm)	~5	15	5–6	2–3	~10
Length (min: um)	1	10	10	1	1
Materials	Si, Ge, GaAs,InP, ZnO, CdSe, GaAs/GaP, GaAs/InAs Si/SiGe	Si, ZnO, Ga_2O_3, t-SnO_2, In_2O_3, CdO, PbO_2, SnO	Ag, cobalt, metals	Group II–IV	Si, InP, ZnO, GaAs
Shape	Vertical; round with blunt tip	Vertical; round with blunt tip	Rod	Vertical; round with blunt tip	Vertical; round with blunt tip
Roughness (optically)	Moderate, smooth	Minor, rough	Rough	Rough	Moderate, smooth
Order	No	No	Yes	No	Yes
Uniformity	No	No	Yes	No	Yes
Throughput	Easier to have bundles than single NW	Low	Easier to have bundles than single NW	High	Easier to have bundles than single NW
Assembling and integration	Difficult	Difficult	Moderate	Difficult	Moderate
Crystalline plane/ orientation	Yes/ {111}	Yes/{111}, {100}	No	Yes/{111}	Yes/{111}, {100}
References	[3, 4, 19–28]	[4, 29–31]	[32–35]	[4, 37–38]	[4, 16, 40–45]

	Top-down approaches						
Dimensions	**Beams lithography**	**Spacer/edge transfer lithography**	**Nano-imprint technology**	**Electro-deposition**	**Oxidation**	**DEA**	**Anisotropic etching**
Width/diameter (nm)	~10	~10	~10	11–30	5	~30	~30
Length (min: um)	1	1	1	1	1	1	1
Materials	Si, Nb, polymer	Si	Si, any materials	Metals	Si only	Si	Si
Shape	Horizontal nanolines	Parallel and horizontal nanolines	Depends on the mold	Line and complex geometry	Depends on initial structure	Rect-angular	Rectangular/triangular
Roughness (optically)	Smooth	Smooth	Smooth	Rough	Smooth	Moderate, smooth	Smooth
Order	Yes	Yes	Yes	Yes	Yes	Yes	Yes
Uniformity	Yes	Yes	Yes	No	Yes	Yes	Yes
Throughput	Slow	High	High	High	High	High	High
Assembling and integration	Easy	Easy	Easy	Moderate	Easy	Easy	Easy
Crystalline plane/orientation	Yes/{111}, {100}	Yes/{111}, {100}	Yes/{111}, {100}	No	Yes/{111}, {100}	Yes/{111}, {100}	Yes/ {111}, {100}
References	[54–55, 58–59]	[57, 61]	[45, 64–65]	[49–53]	[68, 71]	[13]	[12, 69, 75]

In pursuing the highest sensitivity possible for SiNWs as a biosensor, the optimization process is not only on the device fabrication but on the surface modification as well. The parameters between the two different fields should be optimized in an adequate and balanced manner.

NWs are very interesting devices for many more application areas, such as integrated photonics and fuel cells. In all cases, a good and mature fabrication technology is essential to enable the use of NW-based devices to move from a laboratory environment to device production. Choosing or selecting any fabrication approach needs understanding on the application involved and dimension requirements beforehand. Many challenges remain, but we believe that "there is plenty of room at the bottom," as Nobel Prize winner Richard Feynman suggested many years ago.

Acknowledgement

This work is funded by Dutch Technologie Stichting (STW) under project code 10255. The authors would like to thank H.D. Tong and DIMES ICP staff for their advices in cleanroom processing and deep discussions on fabrication of SiNW. LCPMds acknowledges the Netherlands Organization for Scientific Research (NWO) for a VENI grant.

References

[1] G. T. Galyon, "A history of Tin Whisker Theory: 1946–2004," *SMTAI International Conference,* September 26–30 2004.

[2] P. Yang, *et al.*, "Semiconductor nanowire: Whats Next?," *Nano Letters,* vol. 10, pp. 1529–1536.

[3] X. Duan and C. M. Lieber, "General synthesis of compound semiconductor nanowires," *Advanced Materials,* vol. 12, pp. 298–302, 2000.

[4] H. J. Fan, *et al.*, "Semiconductor nanowires: From self-organization to patterned growth," *Small,* vol. 2, pp. 700–717, 2006.

[5] F. Patolsky, *et al.*, "Electrical detection of single viruses," *Proceedings of the National Academy of Sciences of the United States of America,* vol. 101, pp. 14017–14022, 2004.

[6] W. U. Wang, *et al.*, "Label-free detection of small-molecule-protein interactions by using nanowire nanosensors," *Proceedings of the National Academy of Sciences of the United States of America,* vol. 102, pp. 3208–3212, 2005.

[7] K. M. Yasuhide Ohno, Yusuke Yamashiro and Kazuhiko Matsumoto, "Electrolyte-Gated Graphene Field Effect Transistors for Detecting pH and protein adsorption," *Nano Letters,* 2009.

[8] Z. Gao, *et al.*, "Silicon nanowire arrays for ultrasensitive label-free detection of DNA," in *TRANSDUCERS and EUROSENSORS '07 — 4th International Conference on Solid-State Sensors, Actuators and Microsystems*, Lyon, 2007, pp. 2003–2006.

[9] J. I. Hahm and C. M. Lieber, "Direct Ultrasensitive Electrical Detection of DNA and DNA Sequence Variations Using Nanowire Nanosensors," *Nano Letters,* vol. 4, pp. 51–54, 2004.

[10] Z. Li, *et al.*, "Sequence-Specific Label-Free DNA Sensors Based on Silicon Nanowires," *Nano Letters,* vol. 4, pp. 245–247, 2004.

[11] T. Cohen-Karni, *et al.*, "Flexible electrical recording from cells using nanowire transistor arrays," *Proceedings of the National Academy of Sciences of the United States of America,* vol. 106, pp. 7309–7313, 2009.

[12] E. Stern, *et al.*, "Label-free immunodetection with CMOS-compatible semiconducting nanowires," *Nature,* vol. 445, pp. 519–522, 2007.

[13] H. D. Tong, *et al.*, "Novel top-down wafer-scale fabrication of single crystal silicon nanowires," *Nano Letters,* vol. 9, pp. 1015–1022, 2009.

[14] N. E. a. J. Linnros, "Sensitivity of silicon nanowires in biosensor applications," *Journal of Physics,* vol. 100, 2008.

[15] Y. Huang and C. M. Lieber, "Integrated nanoscale electronics and optoelectronics: Exploring nanoscale science and technology through semiconductor nanowires," *Pure and Applied Chemistry,* vol. 76, pp. 2051–2068, 2004.

[16] B. Fuhrmann, *et al.*, "Ordered arrays of silicon nanowires produced by nanosphere lithography and molecular beam epitaxy," *Nano Letters,* vol. 5, pp. 2524–2527, 2005.

[17] Y. Cui, *et al.*, "Doping and electrical transport in silicon nanowires," *Journal of Physical Chemistry B,* vol. 104, pp. 5215–5216, 2000.

[18] C. M. Lieber, "Nanoscale science and technology: Building a big future from small things," *MRS Bulletin,* vol. 28, pp. 486–491, 2003.

[19] F. Qian, *et al.*, "Gallium nitride-based nanowire radial heterostructures for nanophotonics," *Nano Letters,* vol. 4, pp. 1975–1979, 2004.

[20] M. S. Gudiksen, *et al.*, "Growth of nanowire superlattice structures for nanoscale photonics and electronics," *Nature,* vol. 415, pp. 617–620, 2002.

[21] L. Schubert, *et al.*, "Silicon nanowhiskers grown on Si substrates by molecular-beam epitaxy," *Applied Physics Letters,* vol. 84, pp. 4968–4970, 2004.

[22] X. Duan and C. M. Lieber, "Laser-assisted catalytic growth of single crystal GaN nanowires," *Journal of the American Chemical Society,* vol. 122, pp. 188–189, 2000.

[23] S. K. Lee, *et al.*, "Gallium nitride nanowires with a metal initiated metal-organic chemical vapor deposition (MOCVD) approach," *Physica Status Solidi B-Basic Solid State Physics,* vol. 242, pp. 763–763, 2005.

[24] A. Santoni, *et al.*, "Growth of Au-catalysed Si nanowires by low pressure chemical vapour deposition on Si(100) and amorphous Si surfaces," *Journal of Physics Condensed Matter,* vol. 18, pp. 10853–10859, 2006.

[25] L. J. Lauhon, *et al.*, "Epitaxial core-shell and core-multishell nanowire heterostructures," *Nature,* vol. 420, pp. 57–61, 2002.

[26] L. J. Lauhon, *et al.*, "Semiconductor nanowire heterostructures," *Philosophical Transactions of the Royal Society A: Mathematical, Physical and Engineering Sciences,* vol. 362, pp. 1247–1260, 2004.

[27] Y. Wu, *et al.*, "Single-crystal metallic nanowires and metal/semiconductor nanowire heterostructures," *Nature,* vol. 430, pp. 61–65, 2004.

[28] J. Xiang, *et al.*, "Ge/Si nanowire heterostructures as high-performance field-effect transistors," *Nature,* vol. 441, pp. 489–493, 2006.

[29] B. Z. G. Yu D P, Ding Y, Hang Q L, Zhang H Z, Wang J J, Zou Y H, Qian W, Xiong G C, Zhou H T and Feng S Q "Nanoscale silicon wires synthesized using simply physical evaporation," *Appl. Phys. Lett.* , vol. 72, 1998.

[30] Z. R. Dai, *et al.*, "Novel nanostructures of functional oxides synthesized by thermal evaporation," *Advanced Functional Materials,* vol. 13, pp. 9–24, 2003.

[31] R. F. Zhuo, *et al.*, "Morphology-controlled synthesis, growth mechanism, optical and microwave absorption properties of ZnO nanocombs," *Journal of Physics D: Applied Physics,* vol. 41, 2008.

[32] Z. Ye, *et al.*, "'Template-based fabrication of nanowire-nanotube hybrid arrays," *Nanotechnology,* vol. 19, 2008.

[33] M. H. Huang, *et al.*, "Ag nanowire formation within mesoporous silica," *Chemical Communications,* pp. 1063–1064, 2000.

[34] B. Yan, *et al.*, "Fabrication of in situ ultrathin anodic aluminum oxide layers for nanostructuring on silicon substrate," *Applied Physics Letters,* vol. 91, 2007.

[35] Q. Li and R. M. Penner, "Photoconductive cadmium sulfide hemicylindrical shell nanowire ensembles," *Nano Letters,* vol. 5, pp. 1720–1725, 2005.

[36] T. S. Yarden and E. Joselevich, ""Drawing with Nanotubes": Creating Nanowires with Complex Geometries by Pulsed Electrodeposition on Self-Organized Carbon Nanotube Patterns," *Nano Letters,* vol. 10, pp. 4742–4749, 2010.

[37] J. D. Holmes, *et al.*, "Control of thickness and orientation of solution-grown silicon nanowires," *Science,* vol. 287, pp. 1471–1473, 2000.

[38] H. Y. Tuan, *et al.*, "Germanium nanowire synthesis: An example of solid-phase seeded growth with nickel nanocrystals," *Chemistry of Materials*, vol. 17, pp. 5705–5711, 2005.

[39] J. H. Choy, *et al.*, "Soft Solution Route to Directionally Grown ZnO Nanorod Arrays on Si Wafer; Room-Temperature Ultraviolet Laser," *Advanced Materials*, vol. 15, pp. 1911–1914, 2003.

[40] M. C. McAlpine, *et al.*, "Nanoimprint lithography for hybrid plastic electronics," *Nano Letters*, vol. 3, pp. 443–445, 2003.

[41] M. C. McAlpine, *et al.*, "High-performance nanowire electronics and photonics and nanoscale patterning on flexible plastic substrates," *Proceedings of the IEEE*, vol. 93, pp. 1357–1363, 2005.

[42] E. C. Greyson, *et al.*, "Directed growth of ordered arrays of small-diameter ZnO nanowires," *Advanced Materials*, vol. 16, pp. 1348–1352, 2004.

[43] L. Samuelson, *et al.*, "Semiconductor nanowires for 0D and 1D physics and applications," *Physica E-Low-Dimensional Systems & Nanostructures*, vol. 25, pp. 313–318, 2004.

[44] H. T. Ng, *et al.*, "Single crystal nanowire vertical surround-gate field-effect transistor," *Nano Letters*, vol. 4, pp. 1247–1252, 2004.

[45] Y. Zhao, *et al.*, "Fabrication of a silicon oxide stamp by edge lithography reinforced with silicon nitride for nanoimprint lithography," *Journal of Micromechanics and Microengineering*, vol. 18, 2008.

[46] B. J. Ohlsson, *et al.*, "Size-, shape-, and position-controlled GaAs nano-whiskers," *Applied Physics Letters*, vol. 79, pp. 3335–3337, 2001.

[47] Y. Cui, *et al.*, "Diameter-controlled synthesis of single-crystal silicon nanowires," *Applied Physics Letters*, vol. 78, pp. 2214–2216, 2001.

[48] A. I. Hochbaum, *et al.*, "Controlled growth of Si nanowire arrays for device integration," *Nano Letters*, vol. 5, pp. 457–460, 2005.

[49] J. M. Petroski, *et al.*, "Self-assembly of platinum nanoparticles of various size and shape," *Journal of Physical Chemistry A*, vol. 105, pp. 5542–5547, 2001.

[50] C. Xiang, *et al.*, "Lithographically patterned nanowire electrodeposition: A method for patterning electrically continuous metal nanowires on dielectrics," *ACS Nano*, vol. 2, pp. 1939–1949, 2008.

[51] A. Gole and C. J. Murphy, "Seed-mediated synthesis of gold nanorods: Role of the size and nature of the seed," *Chemistry of Materials*, vol. 16, pp. 3633–3640, 2004.

[52] B. D. Busbee, *et al.*, "An improved synthesis of high-aspect-ratio gold nanorods," *Advanced Materials*, vol. 15, pp. 414–416, 2003.

[53] Y. Sun, *et al.*, "Crystalline Silver Nanowires by Soft Solution Processing," *Nano Letters*, vol. 2, pp. 165–168, 2002.

[54] K. Trivedi, *et al.*, "Quantum Confinement Induced Performance Enhancement in Sub-5-nm Lithographic Si Nanowire Transistors," *Nano Letters*, vol. 11, pp. 1412–1417, 2011.

[55] G. C. Tettamanzi, *et al.*, "Superconducting transition in Nbnanowires fabricated using focused ion beam," *Nanotechnology*, vol. 20, 2009.

[56] Z. Gao, *et al.*, "Silicon nanowire arrays for label-free detection of DNA," *Analytical Chemistry*, vol. 79, pp. 3291–3297, 2007.

[57] Y. K. Choi, *et al.*, "Fabrication of sub-10-nm silicon nanowire arrays by size reduction lithography," *Journal of Physical Chemistry B*, vol. 107, pp. 3340–3343, 2003.

[58] H. Fang, *et al.*, "Fabrication of patterned polymer nanowire arrays," *ACS Nano*, vol. 5, pp. 1476–1482, 2011.

[59] R. M. Penner, "Rapid, wafer-scale laser nanoprinting of polymer surfaces," *ACS Nano*, vol. 5, pp. 690–692, 2011.

[60] T. A. Savas, *et al.*, "Properties of large-area nanomagnet arrays with 100 nm period made by interferometric lithography," *Journal of Applied Physics*, vol. 85, pp. 6160–6162, 1999.

[61] Y. K. Choi, *et al.*, "A spacer patterning technology for nanoscale CMOS," *IEEE Transactions on Electron Devices*, vol. 49, pp. 436–441, 2002.

[62] A. Carlson and T. J. K. Liu, "Negative and iterated spacer lithography processes for low variability and ultra-dense integration — art. no. 69240B," *Optical Microlithography Xxi, Pts 1–3*, vol. 6924, pp. B9240–B9240, 2008.

[63] R. Rooyackers, *et al.*, "Doubling or quadrupling MuGFET fin integration scheme with higher pattern fidelity, lower CD variation and higher layout efficiency," *2006 International Electron Devices Meeting, Vols 1 and 2*, pp. 736–739, 2006.

[64] X. M. Yan, *et al.*, "Fabrication of large number density platinum nanowire arrays by size reduction lithography and nanoimprint lithography," *Nano Letters*, vol. 5, pp. 745–748, 2005.

[65] S. Kwon, *et al.*, "Fabrication of metallic nanodots in large-area arrays by Mold-to-Mold Cross Imprinting (MTMCI)," *Nano Letters*, vol. 5, pp. 2557–2562, 2005.

[66] M. Morita, *et al.*, "Growth of native oxide on a silicon surface," *Journal of Applied Physics*, vol. 68, pp. 1272–1281, 1990.

[67] E. Kooi and J. A. Appels, "Selective oxidation of Silicon and its device applications," pp. 860–879, 1973.

[68] G. Pennelli, "Top down fabrication of long silicon nanowire devices by means of lateral oxidation," *Microelectronic Engineering*.

[69] S. Chen, *et al.*, "Top-down fabrication of sub-30 nm monocrystalline silicon nanowires using conventional microfabrication," *ACS Nano*, vol. 3, pp. 3485–3492, 2009.

[70] V. K. Dwivedi, "Bird's-beak-free sealed-interface local oxidation technology for submicron ultra-large-scale integrated circuits," *Journal of the Electrochemical Society*, vol. 137, pp. 2586–2588, 1990.

[71] J. Martinez, *et al.*, "Silicon nanowire transistors with a channel width of 4 nm fabricated by atomic force microscope nanolithography," *Nano Letters*, vol. 8, pp. 3636–3639, 2008.

[72] K. R. Williams and R. S. Muller, "Etch rates for micromachining processing," *Journal of Microelectromechanical Systems*, vol. 5, pp. 256–269, 1996.

[73] J. M. Bustjllo, *et al.*, "Surface micromachining for microelectromechanical systems," *Proceedings of the IEEE*, vol. 86, pp. 1552–1573, 1998.

[74] E. Herr and H. Baltes, "KOH etching of high-index crystal planes in silicon," *Sensors and Actuators: A. Physical*, vol. 31, pp. 283–287, 1992.

[75] T.S.Y. Moh, Y. Maruyama, C. Shen, G. Pandraud, L.C.P.M. de Smet, H.D. Tong, C. van Rijn, E.J.R. Sudhölter and P.M. Sarro., "*IC Compatible Top Down Process for Silicon Nanowire FET Arrays with Three {100} Surfaces for (Bio)chemical Sensing*," *Proceedings of TRANSDUCERS'11*, 2011.

[76] Y. Huang, *et al.*, "Nanowires for integrated multicolor nanophotonics," *Small*, vol. 1, pp. 142–147, 2005.

[77] C. M. Lieber, "One-dimensional nanostructures: Chemistry, physics & applications," *Solid State Communications*, vol. 107, pp. 607–616, 1998.

[78] A. Kargar, "Sensitivity analysis of silicon nanowire chemical sensor," in *2008 8th IEEE Conference on Nanotechnology, IEEE-NANO*, Arlington, TX, 2008, pp. 214–217.

[79] P. R. Nair and M. A. Alam, "Design considerations of silicon nanowire biosensors," *IEEE Transactions on Electron Devices*, vol. 54, pp. 3400–3408, 2007.

[80] R. J. Niklas Elfström, Ilya Sychugov, Torun Engfeldt, Amelie Eriksson Karlström, and Jan Linnros, "Surface Charge Sensitivity of Silicon Nanowires: Size Dependence," *Nano Letters*, vol. 7, pp. 2608–2612, 2007.

[81] B. P. Timko, *et al.*, "Electrical recording from hearts with flexible nanowire device arrays," *Nano letters*, vol. 9, pp. 914–918, 2009.

[82] G. Zheng, *et al.*, "Multiplexed electrical detection of cancer markers with nanowire sensor arrays," *Nature Biotechnology,* vol. 23, pp. 1294–1301, 2005.

[83] R. H. Dennard, "Technology Challenges For Ultrasmall Silicon Mosfet's," *Journal of vacuum science & technology,* vol. 19, pp. 537–539, 1981.

[84] S. Selberherr, *et al.*, "Two dimensional MOS-Transistor Modeling," Urbino, Italy, 1983, pp. 490–582.

[85] Y. Cui and C. M. Lieber, "Functional nanoscale electronic devices assembled using silicon nanowire building blocks," *Science,* vol. 291, pp. 851–853, 2001.

[86] Y. Cui, *et al.*, "Nanowire nanosensors for highly sensitive and selective detection of biological and chemical species," *Science,* vol. 293, pp. 1289–1292, 2001.

[87] G. B. Kang, *et al.*, "Fabrication and characterisation of CMOS compatible silicon nanowire biosensor," *Electronics Letters,* vol. 44, 2008.

[88] X. T. Vu, *et al.*, "Fabrication and application of silicon nanowire transistor arrays for biomolecular detection," *Sensors and Actuators, B: Chemical.*

[89] E. Stern, *et al.*, "Importance of the Debye Screening Length on Nanowire Field Effect Transistor Sensors," *Nano Letters,* vol. 7, pp. 3405–3409, 2007.

[90] P. Offermans, *et al.*, "Gas Detection with Vertical InAs Nanowire Arrays," *Nano Letters,* vol. 10, pp. 2412–2415, 2010.

[91] X. Duan, *et al.*, "Indium phosphide nanowires as building blocks for nanoscale electronic and optoelectronic devices," *Nature,* vol. 409, pp. 66–69, 2001.

Chapter 2

Novel Nanostructured Carbon Nanotube Electron Sources

Matthew T. Cole,* Mark Mann, Chi Li, Kai Hou, Yan Zhang, Xe Qing, Kenneth B. K. Teo, Husnu Emrah Unalan, Wei Lei, Baoping Wang, Daping Chu, Didier Pribat, Pierre Legagneux, Laurent Gangloff, Vincent Semet, Pritesh Hiralal, Gehan A. J. Amaratunga, and William I. Milne*

Department of Engineering, Cambridge University, 9 JJ Thomson Avenue, Cambridge, CB3 OFA, UK

*mtc35@cam.ac.uk, wim@eng.cam.ac.uk

In this chapter, we present a review of our continuing efforts toward the development of discrete, low-dimensional nanostructured carbon-based electron emitters. Carbon nanotubes and nanofibers, herein referred to simply as CNTs, are one-dimensional carbon allotropes formed from cylindrically rolled and nested graphene sheets, have diameters between 1 and 500 nm and lengths of up to several millimeters, and are perfect candidates for field emission (FE) applications. By virtue of their extremely strong sp^2 C–C bonding, intrinsic to the graphene hexagonal lattice, CNTs have demonstrated impressive chemical inertness, unprecedented thermal stabilities, significant resistance to electromigration, and exceptionally high

Advances in Nanodevices and Nanofabrication: Selected Publications from Symposium of Nanodevices and Nanofabrication in ICMAT2011
Edited by Qing Zhang, W. I. Milne, and Jianping Zou

ISBN 978-981-4364-54-6 (Hardcover), 978-981-4364-55-3 (eBook)
www.panstanford.com

axial current carrying capacities, even at elevated temperatures. These near ideal cold cathode electron emitters have incredibly high electric field enhancing aspect ratios combined with virtual point sources of the order of a few nanometers in size. The correct integration and judicious development of suitable FE platforms based on these extraordinary molecules is critical and will ultimately enable enhanced technologies.

This chapter will review some of the more recent platforms, devices and structures developed by our group, as well as our contributions towards the development of industry-scalable technologies for ultra-high-resolution electron microscopy, portable x-ray sources, and flexible environmental lighting technologies.

2.1 Electron Emission

Two dominant forms of electron emission exist; thermionic (or hot) emission, and electric FE. In the more dominant thermionic emission, most often found in cathode ray tubes, metallic filaments are heated to 1000–3000°C, which endows the electrons, in the electron rich metallic emitters, with sufficient energy to overcome the work function of the emitter and emit into the vacuum. Unfortunately thermionic sources are energetically demanding and the high operational temperatures induce severe emitter instabilities and slow on/off times. Contrastingly, FE occurs from a room temperature emitter under the influence of a large electric field. Field emission sources are attractive alternatives to thermionic sources, not only due to their low temperature operation, but also due to their instantaneous response to field variations, high tolerance to temperature fluctuations and incident radiation, high degree of focusability, good on/off ratios, and highly nonlinear emission characteristics [29].

2.1.1 The Field Emission Mechanism and the Fowler–Nordheim Equation

The term *field emission* describes the quantum-mechanical process of electrons tunneling from an electron rich surface, such as a metal, into a vacuum in response to a large electric field [17]. In order to emit electrons at the Fermi level (E_F) must overcome a potential barrier at the surface of the emitter. In the presence of an external electric field, the potential profile, $V(z)$, is modified such that it forms an approximately triangular barrier, as shown in Figure 2.1. For metallic emitters, electron emission occurs when the barrier width is reduced to just a few nanometers. As the field increases, the barrier width decreases, which triggers an increase in the number of tunneling electrons, $J(E)$. The magnitude of this tunneling current can be found by integrating the product of the barrier transmission probability and the incident electron density, $N(E)$, over the range of all possible potential energies. Consequently, the transmission probability depends on the barrier width, the applied field and the work function (ϕ) of the emitter. Assuming a free-electron band structure the emission current density, J (A.cm^{-2}), can be succinctly described by the so-called Fowler–Nordheim (FN) equation, derived via the WKB approximation for emission at metallic surfaces, and is given by [17]:

$$J = \left(\frac{A}{\Phi t^2(z)} \right) E_{\text{local}}^2 \exp\left(\frac{-B\Phi^{3/2} v(z)}{E_{\text{local}}} \right) \tag{2.1}$$

where $A = (e^3/8\pi h) = 1.54 \times 10^{-6}$ A.eV.V^{-2}, and $B = (8\pi/3eh)$ $(2\text{m})^{1/2} = 6.83 \times 10^7$ (eV)$^{-3/2}$V.cm^{-1}, e is the electronic charge, h is Planck's constant, m is the effective electron mass, $t(z)$ and $v(z)$ are the Nordheim ellipse functions, and E_{local} is the local electric field at the emitters apex (V.μm^{-1}), which is equivalent to $\beta.E$, where E is the global electric field and β, the field enhancement factor, is proportional to the aspect ratio of the emitting surface.

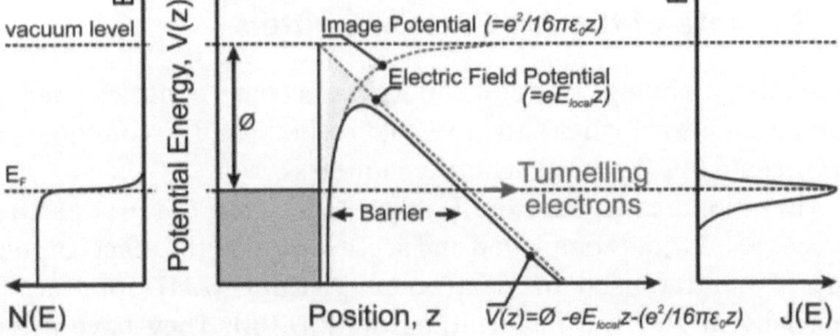

Figure 2.1 Field emission band diagram. Large electric fields induce barrier narrowing which increases the number of electrons tunneling from the metallic, electron-rich, surface into the vacuum. Adapted from [32].

One final and rather elegant simplification was proposed by Brodie *et al.* [8] to modify equation (2.1) into a more tractable descriptor. By substituting the current density and emission area with the measured current, and the local electric field with the product of the applied bias and a cathode geometry dependent field factor, it was found that

$$I = aV^2 \exp(-b/V) \tag{2.2}$$

Here $a = f(\Phi)$ and $b = \beta \cdot \delta$, where δ is the constant anode–electrode separation, are performance metrics that can be empirically determined for any given emitter by measuring the current (I)–voltage (V) characteristics. The turn-on and threshold electric fields are also widely employed as performance metrics. High-performance emitters have low turn-on and low threshold fields. The former is poorly defined and classically refers to the field necessary to emit 1 nA, while the latter refers to the field required to yield a current density of the order of 10 mA.cm^2 [16].

Evidently, in order to extract high, application suitable, currents either E_{local} must be increased or ϕ decreased. In practice a combination of β enhancement and ϕ reduction are employed. For both thermionic and cold-field emitters this is achieved by fabricating sharp tips or columns made from metals like W (4.5 eV) and Mo (4.7 eV), and utilizing low work function alkali metal coatings, such as Cs (1.9 eV) or Ba (2.3 eV). A major drawback here, however, is that most suitably low work function materials have correspondingly low melting temperatures, which give rise to emission instabilities at elevated operational temperatures.

2.1.2 State-of-the-Art Electron Emitters

Commerce leading state-of-the-art electron emitters include lanthanum hexaboride (LaB_6), W cold cathode and zirconium oxide (ZrO)-coated W (ZrO/W) Schottky emitters.

Thermionic LaB_6 emitters (Figure 2.2a) were the first electron sources to be widely employed and are, even today, the most common type of emitter used in electron microscopy (EM) applications, especially in electron beam lithography (EBL). They have a long lifetime (~2000 hours), are very stable and have a low work function (~2.4 eV). LAB_6 emitters are highly processable allowing for application specific tailoring of their geometry and therefore emission characteristics. Conversely, they have a large energy spread (~1.0 eV) and are not particularly bright (10^7 A.cm^{-2}SR^{-1}). The cone angle — the angle subtended at the apex of the tip — is formed through a series of grinding, zone-refining and fine-cutting processes from a larger single LaB_6 crystal. The sharpened apex is then attached to a support by either braising or compression holds between two supporting prongs (Figure 2.2a). For applications such as transmission electron microscopy, where a narrow beam spread is critical, the cone angle subtended is approximately 60°. Contrastingly, for EBL applications, extremely stable sources are required and an angle of 90° is common, resulting in a large emission area, thereby compromising the maximum attainable spatial resolution. LaB_6 emitters are becoming less popular. W thermionic emitters are often opted for despite their substantially shorter lifetime. The main reason for this is that LaB_6 tip processing is rather convoluted, which makes them relatively expensive.

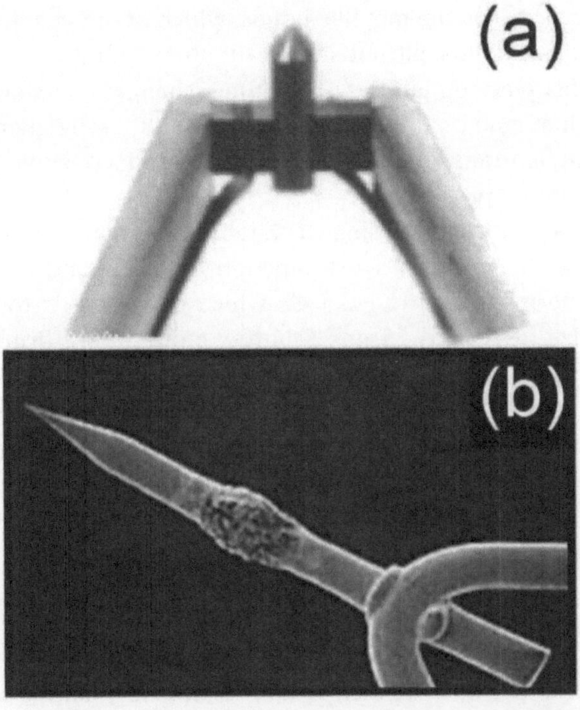

Figure 2.2 Examples of commercially available electron sources. (a) An LAB_6 emitter and (b) an etched W cold cathode emitter.

W thermionic emitters (Figure 2.2b) operate in a similar way to their LAB6 counterparts and have virtually identical functional properties, except for, most importantly, their inferior lifetime (~100 hours). They are, however, extremely reliable and inexpensive. W emitters have relatively large virtual sources (>10^4 atoms) and operate at ~2700°K. Cold cathode W emitters have small virtual source sizes (~3 nm), low temperature functionality (25°C) and life spans in excess of a year. Cold cathode emitters are most often used in high-resolution electron microscopy as they are bright (10^9 A.cm^{-2}SR^{-1}) and have low kinetic energy spreads (0.2 eV). W is the most commonly used material, but it has poor stability. Emission currents can vary by as much as 6% over time. W has a strong affinity toward H_2O, which aids in the processability of the W tips during etching (5 M of NaOH). Unfortunately, H_2O is readily absorbed and lowers the emitter's work function, which induces thermal instabilities. Furthermore, when exposed to the necessary high electric fields the metallic bonds, which are in general weaker than covalent bonds, permit significant atomic diffusion within the W tip. This electromigration, over time, changes the shape of the tip, which augments the electric field profile thereby giving rise to current oscillations, inhibited functionality, and a positive-feedback tip degradation process.

Schottky emitters are single crystal W tips coated with ZrO. The ZrO lowers the surfaces work function to 2.3 eV and reduces the kinetic energy spread to ~0.3 eV, which corresponds to a virtual source size of around 15 nm. Schottky emitters are brighter (10^8 A.cm^{-2}SR^{-1}) than both W and LAB_6 emitters, and have a typical stability of <1%. Substantial lifetimes are also common (1–2 years). Unfortunately the noise level of Schottky emitters tends to increase with time. Interestingly, however, Schottky emitters have the best lifetime characteristics of all emitters. The deposited Zr readily evaporates when exposed to vacuums. However, it is rapidly replaced from the Zr reservoir within the W bulk, which gives rise to its stable lifetime characteristics.

Cold cathode, thermionic and ZrO-coated Schottky sources are fabricated from chemically etched monocrystalline W wires, typically 100 μm diameter and around 5 mm in length, spot-welded to W crosswires of similar dimensions that are attached to thick covar prongs (Figure 2.2b). Of the total throughput manufactured by FEI approximately 70 % are considered suitable for sale.

Si has also been widely considered as an attractive material for FE applications due to the highly developed global processing infrastructure. However, the demanding fabrication processes necessary to form the relatively low-aspect ratio tips and Spindts has severely inhibited the dominance of Si-based electron emitters.

EM has the highest FE requirements, and as such provides the most appropriate benchmark with which to compare the pervading technologies. EM applications demand bright, stable, low-noise electron sources with low kinetic energy spreads to maximize spatial resolution and image contrast. Recent research, performed by us and others, investigated the appropriateness of CNTs as electron emitters for such applications and, perhaps more importantly, their performance compared with the state-of-the-art commercially available technologies. The key performance metrics are summarized in Table 2.1. To better understand the requirements of an ideal electron source, and how CNTs perform with respect to this idealized technology, the key parameters from Table 2.1 have been normalized to this conceptual optimally performing technology and illustrated in the figure-of-merit polar plot appended to Table 2.1. The circumferential pink-dashed line marks the ideal scenario, a technology that has; a small virtual source, a low kinetic energy spread, high brightness, high stability, a high lifetime, a low work function, a low operating temperature, high operating vacuum, a low extraction electric field, and a high current density. In all metrics CNT-based electron sources perform extraordinarily well compared with the dominant technologies previously discussed. CNT-based field emitters are still an emerging technology that accounts for their relatively low-level of stability thus far attained. However, it is important to note that the measured stabilities are still comparable to commercially available thermionic (W and LAB_6) and Schottky emitters and it is predicted they will soon exceed these technologies.

Table 2.1 Typical operating parameters* of the various industry pervading electron sources (Therm.(ionic), Schot.(tky), and C.C.-Cold cathode) compared with CNT field emitters

	Therm. W	Therm. LAB_6	Schot. W/ZrO	C.C. W	CNTs
d_s (nm)	10,000	10,000	20	3	10
D (nm)	60,000	10,000	<1,000	<100	<50
ΔE (eV)	1.5	1.0	0.7	0.2	0.3
β	10^6	10^7	10^8	10^9	10^9
I_{max} (nA)	10^3	10^3	10^2	0.2	5×10^3
τ (Yr)	0.1	0.2	>1	>1	1
Φ (eV)	~4.5	~2.7	~2.8	~4.5	~5.0
T (°C)	~2200	1500	~1700	25	25–400
Stability (%)	<1	<1	<1	5	0.5
J_c (A.cm^{-2})	1	10^2	10^3	10^5	10^7
Op. Vac. (mbar)	10^{-7}	10^{-8}	10^{-9}	10^{-10}	10^{-7}
E (V.µm^{-1})	10^1	10^1	10^2	10^3	5

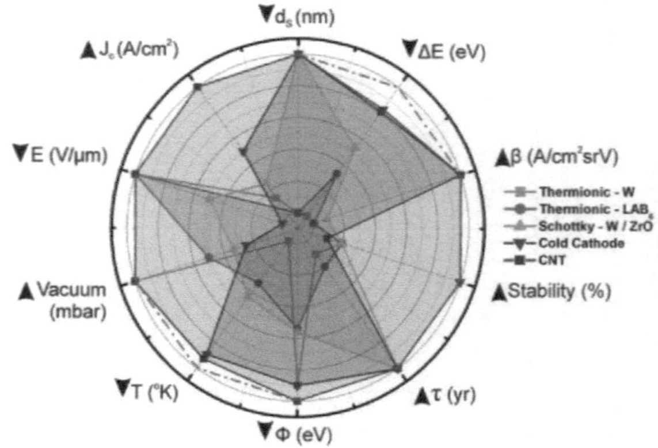

Note: As seen in the figure-of-merit polar plot, CNT-based field emitters are seen to outperform all of the state-of-the-art electron sources across most metrics.

* d_s the virtual source size, D is the cathode radius, ΔE the energy spread of the emitted electronics, β the electron-optical brightness, I_{max} the maximum probe current, τ the lifetime, Φ the work function, T the typical operating temperature, J_c the current density, Op. Vac. is the operating vacuum and E is the turn-on electric field.

2.1.3 The History of Carbon-Based Field Emitters

One of the earliest reports of carbon-based FE was by Baker *et al.* [3] in 1972 who noticed that, in various environments, graphite fibers showed enhanced emission stability compared with several common metallic emitters. Wang *et al.* [48] and Geis *et al.* [18], in 1991, using diamond-based cathodes, reported electron emission at surprisingly low threshold fields (<3 $V.\mu m^{-1}$). Later that same year, Djuba *et al.* [15] demonstrated that arrays of diamond-like carbon (DLC) cones emitted at extraction potentials much less than those necessary for Mo or Hf tips. It was these, arguably seminal works that stimulated more extensive research into the potential for high-performing FE from carbon-based materials.

It became apparent that field emitting carbon materials could be classified into two distinct groups based on their proposed emission mechanism. These were (i) the structured graphites (sp^2) where the emission was a result of field enhancement associated with the geometry of the conductive emitter, and (ii) the diamond types (sp^3) where the negative electron affinity [4, 22, 47] and band bending effects [19, 20] were important. The belief, at the time, behind the use of wide band gap materials, such as diamond and its synthetic derivatives, was that the electron affinity was low, or even negative in some cases. Thus, electrons injected into the conduction band observe a reduced potential barrier and, as a result, easily emit into vacuum. It is now accepted that this is not the case. Such band bending should, in principle, apply homogenously over the entire surface resulting in spatially uniform emission, and yet the observed FE always occurred at a few highly localized and randomly distributed sites. The widely accepted model and data were inconsistent. Spatial mapping of the FE characteristics hinted that there was something peculiar about the emission sites in relation to the rest of the planar carbon coatings [26].

In the case of chemical vapor deposited (CVD) diamond coatings, simultaneous photoemission and FE studies showed that the excited electrons were sourced from ~5–6 eV below the vacuum level [26]. It soon became clear that the field enhancing characteristics originated from the graphitic grain boundaries [39], a result that was largely supported by the fact that highly defective and interface-rich films exhibited superior FE properties compared with decidedly planar films. As a result, tetrahedral amorphous carbon (*ta*-C), a

DLC derivative, was heavily studied. It appeared that the *ta*-C films only ever emitted after *activation*. That is, when a vacuum arc discharge event occured with sufficient vigor to perturb the surface topography to form a tip-like sites, typically at applied fields of 80–180 V.μm^{-1}. Post FE scanning electron micrograph EM analysis of these *ta*-C coatings revealed sharp topographic features, which had a propensity to function as suitable geometrically enhanced emission sites [41]. Prior to FE activation no surface perturbations existed and poor emission characteristics were recorded. Further analysis showed that what were originally believed to be smooth *ta*-C films where, in fact, covered in sub-micron-sized graphitic inclusions and particles [45], which function as appropriate emission sites with $\beta \sim 250$ — field enhancement factors far too high to be consistent with the films planar nature.

Consequently a new trend emerged. The production of intentionally diverse composite carbon films, rich in a variety of graphitic nanostructures, became commonplace. These films contained incredible assortments of naturally occurring carbonaceous nanostructures, including, but certainly not limited to; particles, onions, clusters, fullerenes, nanotubes and nanofibers. These nanostructured carbon films performed significantly better than the carbon leading materials at the time — *ta*-C and DLC — with low turn-on fields of the order of 1–5 V.μm^{-1}.

More recently, FE has been reported from artificial CVD diamond [51], *a*-DLC [2, 37], CVD graphite [34], and various polymers [31]. However, out of those materials, those with the highest aspect ratios or particularly perturbed surfaces were found to outperform the more conformal and smooth coatings. Evidently sharp tips were still preferential, even in the case of carbon-based emitters. Thus, the stage was set for the dramatic insertion and intense interest in CNT field emitters.

2.1.4 Carbon Nanotube and Nanofiber-Based Field Emitters

CNTs are formed from individual and nested rolls of graphene, a single atomic plane of hexagonally latticed carbon atoms (Figure 2.3). A holistic heterogenous catalysis mechanism has not yet been fully developed. However, some consistencies exist between studies. Upon thermal stimulation the transition metal catalyst (typically Fe, Ni or Co) agglomerate in a process known as Oswald ripening, an effect driven by free surface energy minimization, to form a catalyst bed that consists of many millions of nanometer-sized spherical and hemispherical catalyst islands. The detailed size, distribution and phase of these islands depends critically on the amount of catalyst and barrier-layer material employed, as well as the annealing temperature and annealing atmosphere. These nanoisland catalysts are then exposed to a carbon containing gas, typically acetylene (C_2H_2) or methane (CH_4). The carbon containing gas undergoes pyrolysis and the catalyst nanoislands become supersaturated with the liberated carbon. This carbon is then extruded from the catalyst particle to form the CNT.

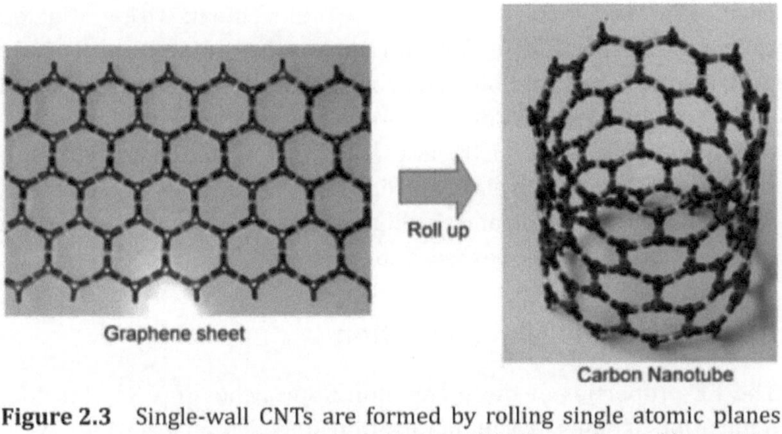

Figure 2.3 Single-wall CNTs are formed by rolling single atomic planes of graphite (graphene). Multiwall CNTs are formed by nesting multiple cylinders [29].

To-date most CNT-based field emitters have been fabricated by depositing independently mass produced CNT-containing pastes or powders onto various substrates. The source CNT material is most often synthesized by arc-discharge, thermal and plasma enhanced CVD, ball milling, or laser ablation. Regardless of the exact synthesis process, the source CNTs are then purified and mixed with an epoxy or binder and are either screen printed, cast or sprayed to form large area films or selectively located using EM-monitored nanomanipulators [12]. Electrophoresis has also been exploited to suspend CNTs at specific locations [10]. The emission characteristics of various CNTs types and geometries has been investigate, including individual single-wall [14] and multiwall CNTs [36], CNT networks embedded in epoxy matrices [40], and vertically and horizontally aligned arrays [23, 29, 30].

2.2 CNT Thin Films on Flexible Substrates

For high throughput, low-end applications, such as environmental lighting and flexible FE-based display backlighting, simple low-cost processing is a stringent requirement for successful commercialization. Functionality at comparatively high pressures (10^{-6} mbar) with currents of the order of 10 mA is also necessary. The potential advantages of mechanical flexibility and large-area processing are also clear. To meet these requirements we investigated various CNT thin film emitters, rich in electron emitting surface protrusions, fabricated on polymeric substrates. Quartz supported CNT thin film electron sources, providing a current density of 100 $mA.cm^{-2}$, are commercially available in portable X-ray systems. However, flexible FE displays and lighting are not. To appreciate why, some understanding of, and the limitations associated with, the thin film deposition techniques were considered.

2.2.1 CNT Thin Film Deposition

The FE properties of three common approaches to CNT thin film deposition (rolling, vacuum filtration and screen printing) were investigated (Figure 2.4). The source material for the rolled CNT film was synthesized by thermal CVD on a Si/SiO_2 (200 nm) substrate coated with a bilayer catalyst of magnetron sputtered Al2Ox (10 nm) and evaporated Fe (1 nm) in a 4% high purity H_2 diluted C_2H_2 atmosphere at a pre-pressurized chamber pressure of 26 mbar heated to 700°C for 10 minutes. Vertically aligned (500 μm tall, 40 nm diameter) MWCNTs were formed. Flexible field emitters were fabricated by rolling the source MWCNTs onto ITO-coated 150 μm polyethylene terephthalate (PET) substrates pre-treated with a self-assembled monolayer adhesion promoter: aminopropyl triethoxysilane (APTES). Transferred MWCNT films were ~20 μm deep (Figure 2.4a,b).

For the vacuum filtration emitters, 2 mg.L^{-1} MWCNTs were cleaved from the source MWCNT substrate and dispersed in a solution of de-ionized water and sodium dodecylbenzene sulfonate (1 L:700 mg), using a high power tip ultrasonicator (250 W / 24 kHz) for 10 h. The dispersion was filtered through a 50 μm PET mesh to remove residual amalgamates. This solution was then vacuum filtrated through a mixed cellulose ester membrane (0.22 μm pore) and rinsed with de-ionized water. This MWCNT/MCE composite was placed in contact with an Al-coated glass substrate and heated to 80°C in ambient under a compression of 35 g.cm^{-2} for 3 h. Once dry, the MCE was gently peeled away to activate the film and the remaining emitter was rinsed in acetone, IPA and de-ionized water and dried with N_2 (Figure 2.4c,d). The emitter had a 550 nm optical transparency of ~40% and a sheet resistance of 350 Ω.□$^{-1}$.

A third CNT field emitter was fabricated by screen printing a commercially available (Xinano, XPP10) 8.3% CNT concentration (60 %wt. solid content) paste onto a screen printed Ag paste electrode (sintered at ~560°C) on Corning glass substrates (Figure 2.4e,f). Samples were sintered at ~520°C in an N_2 atmosphere to increase the adhesion between the Ag paste electrode and MWCNT film and to remove any remaining organic binders. Figure 2.4 shows SEM micrographs of the prepared emitters. Al/glass and ITO/PET controls were prepared by magnetron sputtering. The cathode was fabricated by spray coating a 1:20 (by wt.) polymethylmethacrylate (PMMA):phosphor mixture on Corning glass. A 30 nm Al top contact was deposited by magnetron sputtering forming the partially transparent scintillator.

The *I–E* plot and FN characteristics of the three emitters are illustrated in Figure 2.4g,h. The FN equation is typically applied to metals and predicts no saturation due to high conductivities and high electron densities. However, in lower conductivity films internal resistance and contact resistance must be considered, especially under high fields. Saturation currents of 2.6, 1.5, and 1.3 mA were observed at applied fields of 1.5, 2.5, 2.6 V.μm^{-1} for the rolled, vacuum filtered and screen printed networks, respectively. In the low-field regime (<1.5 V.μm^{-1}), conventional linear FN characteristics were observed. However, the emission characteristics deviate from this in the high-field region due to saturation effects. Similar effects were observed in all three cases. Collins *et al.* [11] attributed this deviation from typical FN behavior to the electrostatic interaction between neighboring CNTs. However, this saturation effect is more likely ascribed to the internal resistance of the MWCNT film (i.e. the inter-CNT tunnel-barrier contact resistance) and the contact resistance between the MWCNT film and the underlying conductive substrate. Moreover, the interaction between the MWCNT dipoles and the electric field dynamically adjusts the nanoscopic structure of the film, increasing the mean inter-tube separation. In percolation networks, such as these, conduction is largely dominated by variable range hopping [6]. Under increasing electric fields the inter-tube separation increases and the effective tunnel barriers between CNTs increases. As the barrier width increases, the sheet resistance correspondingly increases, which reduces the maximum emission current. The dynamic nature of the film ballasts the emission characteristics. However, this dynamic current reduction is offset by an increase in local field enhancement due to topographic variations associated with the nanoscopic motion in the film. Individual tips move and start to dominate the emission. Poorly bound CNT films, such as the rolled film, move more freely than bound CNTs, such as the screen printed emitters, causing saturation at lower fields. Interestingly, however, we observe higher maximum emission currents for the rolled films as the topographic alterations (which increases the number of emission sites) dominates the current reduction by enhanced inter-tube tunnel barriers.

Two distinct β-regimes are evident in the FN plots of Figure 2.4h. For CNT-based thin films these two slopes are attributed to the switchover from a predominately tip emission to side emission, resulting from the differences in interface barriers and emission geometries. This effect is consistent with the data presented by [38] and accounts for the lack of a secondary β-regime in the case of vertically aligned CNT arrays. Many factors influence the measured β factors. Rolled films possess drastically smoother surface topography. The RMS roughness/thickness ratios of the screen printed and vacuum filtrated films are comparable (2.5 and 0.6, respectively). However, this is not the case for the rolled thin films (132.4). For a ratio >10, films are considered smooth as the film roughness is much less than the film thickness resulting in a response more characteristic of the bulk material (i.e. a low β). For a ratio <1 films are very rough and one would expect high field enhancement.

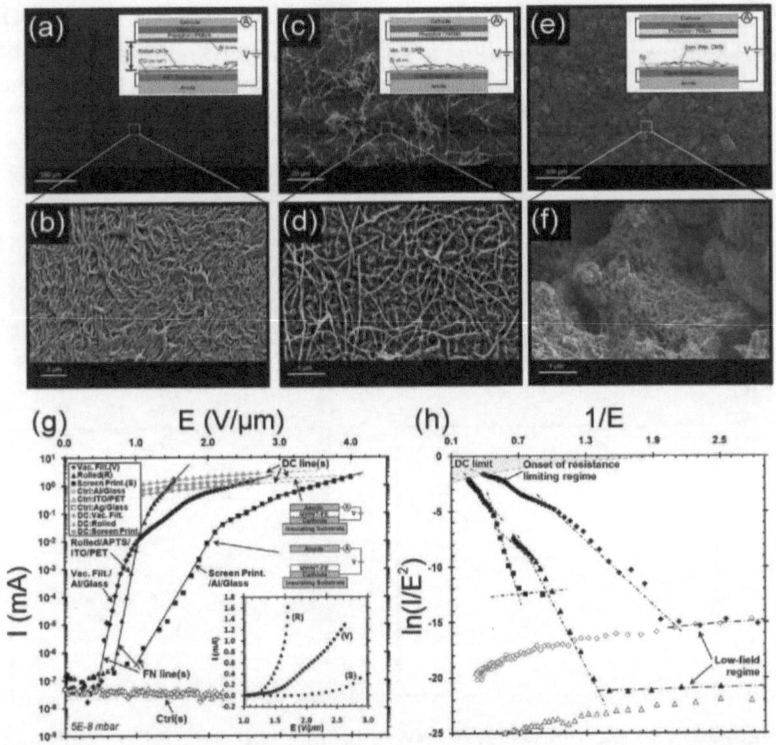

Figure 2.4 SEM micrographs of thin film CNT-based emitters: **(a, b)** Rolled **(c, d)** Vacuum filtrated **(e, f)** Screen printed. **(g)** The electric field dependence of the emission current. *Lower right inset*: *I–E* data. *Upper right inset:* Measurement setup employed to distinguish the FN regime and the resistive DC limit. **(f)** Corresponding FN plot.

2.2.2 Ballasted CNT/ZnO Nanowire Thin Films

In an attempt to limit the electron emission from dominant tips within the deposited CNT films, zinc oxide (ZnO) nanowires (NWs) were deposited on top of the films to form ballast resistor-like structures (Figure 2.5). The ZnO NWs function as equivalent series resistances that limit the emission current of emission-dominating CNTs, while simultaneously encouraging emission from other CNTs. Areal SEM and optical micrographs of a rolled CNT film and coated ZnO NW film are shown in Figure 2.5a–c. Figure 2.5d shows an optical micrograph of a ZnO

NW-coated CNT thin film. The emission characteristics of the CNT and the ZnO NW+CNT films are shown in Figure 2.5e. Figure 2.5f illustrates the FN plot. The ZnO NWs can be seen to reduce the maximum emission current by around one order of magnitude.

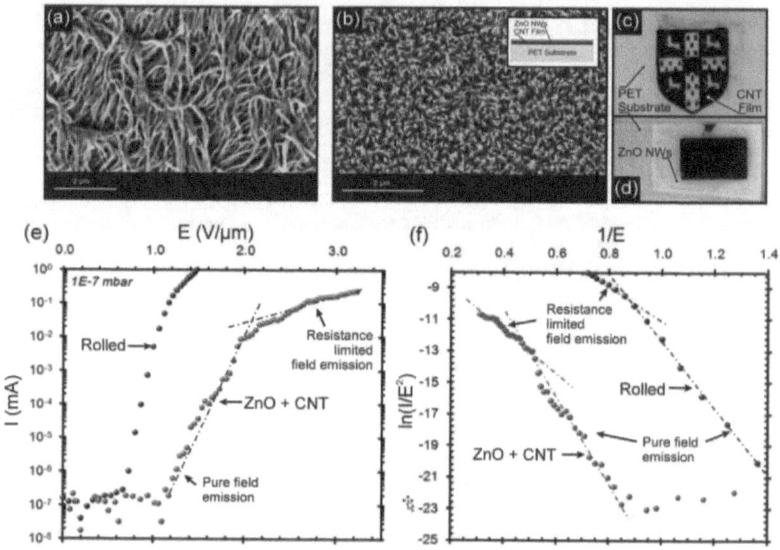

Figure 2.5 Ballasted CNT/ZnO NW thin films. **(a)** SEM micrograph of a rolled CNT thin film, and **(b)** ZnO NW-coated CNT film. *Inset:* Schematic cross-section of the flexible emitter structures. Optical micrographs of **(c)** a patterned CNT emitter and **(d)** a ZnO NW-coated CNT emitter on APTES pre-treated PET substrates. **(e)** *J–E* characteristics and **(f)** the corresponding FN plot.

Although we have repeatedly demonstrated that high currents are attainable using carbon-based thin film emitters, high current densities are not. Vertically aligned CNT arrays solve this problem. We have deposited highly reproducible periodic CNT arrays, >1 cm^2 in size, with emission site densities as high as 2.5×10^7 cm^{-2}. To increase the emission current densities further, which are necessary for x-ray and microwave applications for example, much higher β factors are necessary. For thin film technologies high packing densities within the films results in significant electrostatic shielding, which significantly reduces the effective β. The spatial uniformity of the emission sites is also particularly poor leading to "hot spot" formation. Groening *et al.* [21] and Bonard *et al.* [7] determined that close packing between CNTs electrostatically screens the applied field and effectively reduces the field enhancement factor. Figure 2.7a illustrates this nearest neighbor electrostatic shielding process [29]. Lines of equipotential illustrate that although the composite CNTs have high aspect ratios (and therefore potentially high field enhancement factors), it is not possible to access them. As a result, the current densities are substantially reduced. The inter-CNT pitch must be approximately twice their height in order to minimize nearest-neighbor shielding for maximized emission current (Figure 2.7b) [29]. Accurate control over the CNTs alignment was also found to be critically important in optimizing the FE properties. Plasma enhanced CVD (PE-CVD), combined with EBL and nanoimprint lithography, offers a viable means of accurately controlling not only the CNT position but also the alignment, length and even the crystallographic structure.

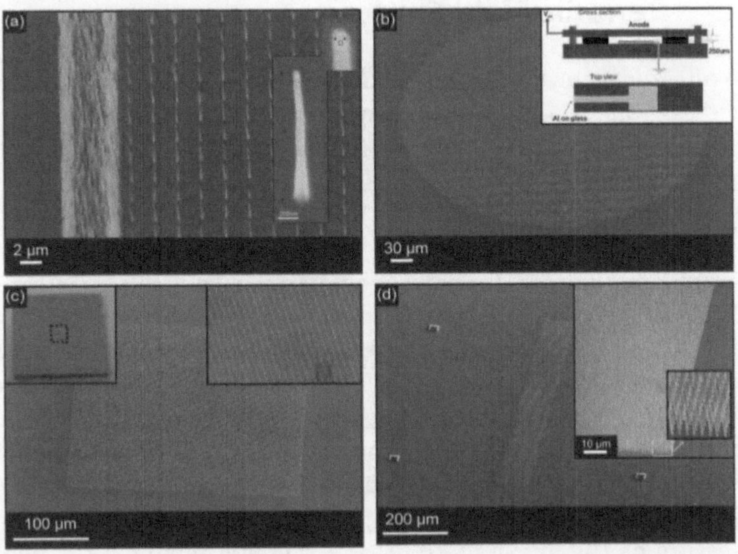

Figure 2.6 SEM micrographs of various PE-CVD synthesized periodic CNT arrays fabricated in our group. **(a)** An encapsulated square CNT array. *Inset:* High- magnification of an individual CNT. **(b)** A circular array. *Inset:* Measurement setup. **(c)** Square array. *Top left inset:* Optical micrograph of a 5 × 5 mm chip with the dot array area highlighted. *Top right inset*: High-magnification micrograph of the CNT array. **(d)** Rectangular array. *Top right inset:* High-magnification micrographs of CNTs in the rectangular array.

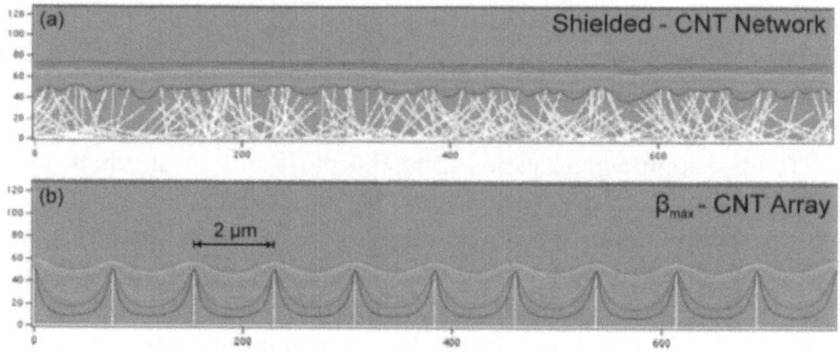

Figure 2.7 Simulated electrostatic screening interactions. **(a)** Significant screening occurring in dense CNT films, and **(b)** maximized CNT packing density simultaneously achieving high current densities with minimized shielding effects.

2.3 Periodic Field Emitters

Engineered vertically aligned periodic CNT arrays [43] simultaneously increase uniformity, field enhancement, and emission currents compared with random CNT thin film networks.

2.3.1 Periodic Arrays

Various periodic CNT array geometries have been considered. The fabrication process is largely the same for each, and has been reported in detail elsewhere [38]. CNT arrays were synthesized on degenerately doped *n*-type Si in a custom-built plasma enhanced (PE)-CVD system. EBL and PMMA lift-off, combined with magnetron sputtering, were employed to produce arrays of 7nm-Ni catalyst dots, 100 nm in diameter and 6 µm pitch. The substrate, placed on a resistively heated graphite stage, is then thermally ramped, at a pressure of 10^{-2} mbar, to 700°C at a ramping rate of 100°C.min^{-1}. During heating, NH_3 is introduced at a fixed chamber pressure of 2.4 mbar. C_2H_2 is then introduced once the temperature reached 750°C. A low-power direct current plasma assists synthesis and aligns the CNTs.

Based on the estimated upper current emission capabilities of individual CNTs, circular patterns 300 µm in diameter containing approximately 10^4 CNTs with a pitch of 2.5 µm, were fabricated. The suitability of rectangular (100 µm × 1 mm, 1.1×10^4 emitters, 1.5 µm in height, 3 μm pitch) and square (300 μm sides, 1.44×10^4 CNTs, 2.5 μm pitch, 1.25 μm in height) arrays was also investigated. Figure 2.6 shows SEM micrographs depicting the fabricated CNT arrays. All the arrays exhibited excellent height uniformity over large areas. The inset of Figure 2.6b shows a schematic of the FE measurement setup. Using periodic arrays such as these we have obtained current densities of the order of 1.0 $A.cm^{-2}$ operated in DC mode, and >12.0 $A.cm^{-2}$ in pulsed mode (20% duty cycle) at RF frequencies [44]. Although the CNT arrays exhibited excellent height, diameter and crystallographic uniformity, which leads to high current densities, improvement in the temporal stability is necessary as, in spite of the impressive functionality the arrays demonstrated tendencies to over emit through specific CNTs resulting in emitter burn-out, an effect compounded by localized ion bombardment associated with residual gas species and trapped gas volumes. To solve this problem ballast resistors were integrated with these highly regular engineered structures, in much the same way as discussed previously. However, in this case the use of current-limiting resistances in series with *each* CNT was investigated to ensure current contributions are efficiently redistributed across all the emitters.

2.3.2 SOI Substrate Ballasted CNT Arrays [52]

As in the case of the CNT thin film emitters discussed previously, in order to attain large current densities with reasonable uniformities, ultra stabilities and long lifetimes, it is necessary to prevent particular CNTs in the arrays from dominating the emission. Thus, we investigated the possibility of integrating ballast resistances for each individual CNT. A typical ballasted array is shown in Figure 2.8. Here CNT arrays, grown on Si-on-insulator (SOI) substrates (Si/SiO_2/Si), were electrically connected through a controllable resistor in the form of the undoped Si between the W biasing electrode and the CNT emitter. This Si gap functions as the controllable current-limiting in-plane ballast resistance. As before, vertically aligned CNT arrays (50 nm in diameter, 10 μm pitch) were deposited through a combination of EBL patterning, Ni catalyst deposition and PE-CVD synthesis in an NH_3:C_2H_2 atmosphere at temperatures of ~700°C. Between the catalyst deposition and CNT growth, W biasing electrodes were EBL defined and magnetron sputtered. The size of each grid is 2.5 μm × 2.5 μm. The W electrodes were 0.5 μm wide (Figure 2.8a). Individual CNT emitters are located at the centre of each grid, and have a nominal length of 1.25 μm (upper right inset Figure 2.8a). The lower left inset of Figure 2.8a shows a schematic diagram of the measurement system.

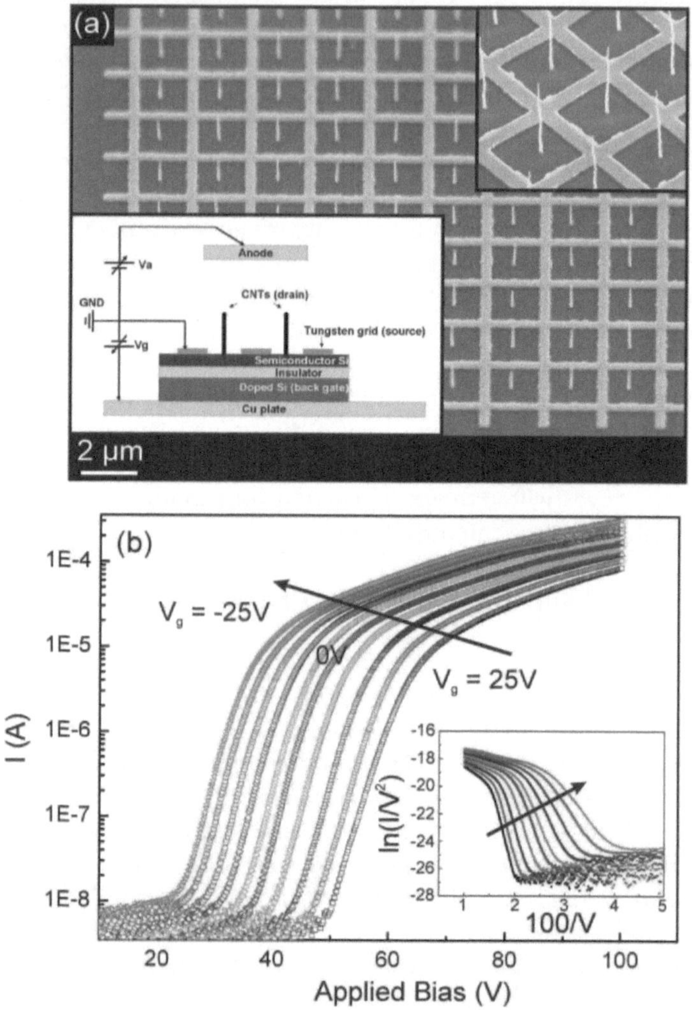

Figure 2.8 SOI substrate-ballasted CNT array. **(a)** SEM micrograph of a substrate ballasted SOI-CNT field emitter array. *Lower left inset:* Schematic showing the measurement setup. *Upper right inset:* High magnification SEM micrograph of six CNT emitters and the W source electrode. **(b)** Measured emission current as a function of the anodes applied extraction bias. *Inset:* Corresponding FN plot.

The ballasted structure operates in a transistor-like fashion. The W grid functions as the source electrode, the CNT as the drain electrode, and the underlying Si substrate (*not* the ballast Si) as the channel modulating gate electrode that electrically modulates the conductivity of the Si channel.

Figure 2.8b shows the variation in emission characteristics as a function of the gate voltage and the consequent FN plot. The observed characteristics are a result of two competing trends, at a fixed extraction potential. The Si ballast resistance reduces the potential at the CNT and therefore increases the extraction potential difference (between the extraction anode and the CNT emitter), which increases the emission current. However, this is offset by the ballast resistance limiting the number of electrons that can be emitted. This second process dominates. Although an increased emission current is accessible, by virtue of the maximized extraction electric field, only electrons in the comparatively electron deficient CNTs are accessible, thus the emission current is severely inhibited by the ballast resistance. The CNT emitters can only emit electrons, not source them. Electron sourcing must come from the W drain electrodes through the ballast Si channel. As the emission current increases, the potential at the drain contact becomes increasingly positive and the gate-drain potential difference is reduced leading to a pinch-off-type current saturation. The devices exhibited *p*-type behavior with an on/off ratio of $\sim 10^4$.

Efficient emission current control was achieved *via* the intrinsic feedback signal. As the emission current increases, in response to an increasing anode-CNT cathode potential difference, the potential drop across the channel of the transistor also increases, tending to limit the rise in emission current by reducing the tip-to-anode potential difference.

Electrons accumulating at the drain (CNT) of the transistor are "hot" due to the potential drop across the Si channel. These hot electrons efficiently propagate through the CNT and are readily field emitted. In a conventional FE system, the electron energy distribution is characterized by the ambient temperature and defects along the length of the CNT that scatter the electrons being transmitted towards the tip, ensuring their thermalization. In this case, FE is determined by the transmission probability of the electrons from the Fermi Sea in the tip of the CNT, through the triangular potential barrier, into the vacuum. The injection of hot electrons from the ballast transistor modifies the transport through the CNT. These electrons are scattered much less effectively and arrive at the tip with sufficient excess momentum to significantly enhance the transmission into the vacuum.

The effect of the ballast transistor is to increase the current contribution of every tip in the low-field region and to limit the current in the high-field region due to the resistance of the channel. Combining these effects improves the emission uniformity and lifetime as reduced electric fields are required for a given total current.

The main problem associated with this ballasted-structure was that the W source electrodes limit the attainable CNT packing density, and therefore the maximum emission current. The W bus also imparts undesirable beam shielding and focusing effects. Furthermore, reducing the number of patterning steps, especially in the case of EBL, increases the commercialization potential of any technology. Thus, in an attempt to remove the costly and time-consuming patterning processes necessary, we investigated the possibility of directly coating CNT arrays with ballast resistor-like structures without compromising their field enhancing characteristics.

2.3.3 ZnO NW Ballasted CNT Arrays

CNT arrays, such as the one shown in Figure 2.6, were grown and ZnO NWs were deposited directly onto them using a hydrothermal approach [46]. Zinc acetate dehydrate in 1-propanol (spectroscopic grade) was spin coated onto the vertically aligned CNTs at 2000 rpm for 30s. The substrates were then annealed at 100°C for 2 minutes to promote adhesion. A uniform seed layer was obtained after three casts. Prepared substrates were then dipped into an equimolar solution of 25 nM zinc nitrate hexahydrate ($Zn(NO_3)_2{\cdot}6H_2O$) and hexamethylenetetramine in de-ionized water heated to 80°C in order to synthesis the ZnO NWs. Although still requiring a single EBL or nanoimprint step, an important benefit of these structures compared with the substrate ballasted emitters (discussed above) was the less demanding fabrication process [24].

A comparison between the emission characteristics of a standard CNT array and the CNT/ZnO NW-coated array is shown in Figure 2.9b,c [24]. The unexpectedly low turn-on field for the ZnO/CNT emitters is due to the increased field enhancement caused by the addition of the high-aspect ratio ZnO NWs. The average field enhancement

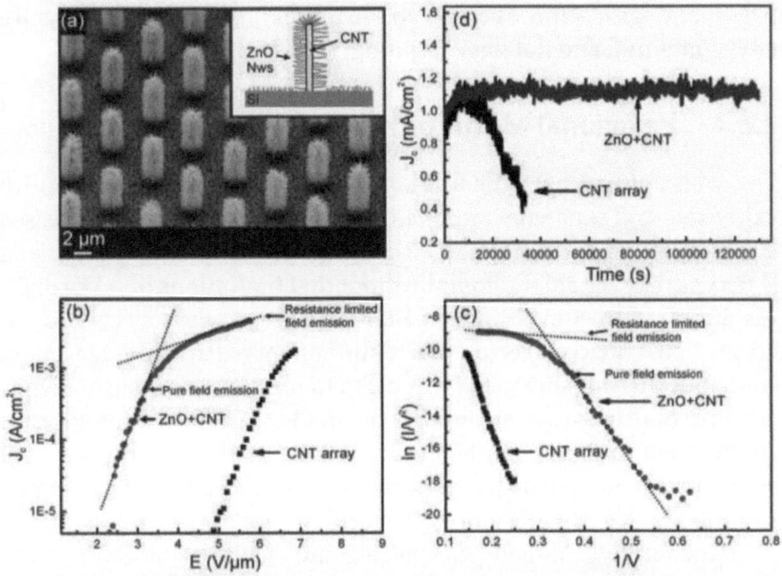

Figure 2.9 ZnO NW ballasted CNT arrays (ZnO+CNT). **(a)** SEM micrograph of ZnO NW-coated CNT array. *Upper left inset:* Schematic cross-section of a coated CNT. **(b)** *J–E* comparative plot and **(c)** corresponding FN plot for a standard CNT array and a ZnO NW-coated CNT array. **(d)** Temporal stability comparison between the raw CNT array and the ZnO NW-coated CNT array [24].

factors of the CNT array and CNT/ZnO were calculated to be 820 and 2400, respectively [24]. The emission current densities were a few $mA.cm^{-2}$ in both cases. The ZnO NWs tended to reduce the maximum emission current by approximately a factor of two – as was the case in the ZnO NW-coated CNT thin films discussed previously.

One of the key effects of the ballasting ZnO NWs was to improve the lifetime and stability of the array - the key barrier to the successful development of FE-based lighting devices. We have, thus far, conducted accelerated lifetime measurements at a pressure of 5×10^{-6} mbar with an initial emission current density of 1 $mA.cm^{-2}$ [24]. Figure 2.9d shows the emission current density as a function of time. The ZnO-coated arrays show a substantially enhanced lifetime, but little improvement in stability (i.e. noise margin). The limited carrier concentration in the ZnO restricts the emission current thereby protecting the underlying CNT emitters from high pressure degradation, opening up the possibility of modest pressure environmental and flat panel lighting, for example.

2.3.4 Hexagonal Mesh Arrays

The well-defined periodic arrays discussed at some stage, require expensive and time-consuming EBL patterning. Such EBL processes substantially reduce commercial competiveness. Simpler processing is necessary to avoid the complications and limitations of serial high-resolution patterning. Efficient FE is known to occur at CNT forest edges and vertices. Vertex-rich emitters increase the edge effect contributions and enhance the local field profiles resulting in higher current densities at lower electric fields. COMSOL multiphysics simulations showed that the electric field distribution for for many geometries, but particularly hexagonal meshes, is indeed highly confined to the edges (Figure 2.10a) [9].

Degenerately doped *n*-type Si and glass substrates were patterned using a simple low-cost parallel imprint lithography process to form 1-μm-wide Mo electrodes in a tessellated hexagonal array. Bilayer catalyst structures were then deposited (magnetron sputtered; natively oxidized 10-nm-Al / 1-nm Fe). Vertically aligned CNT forests were then synthesized by thermal CVD by annealing in an Ar atmosphere at 420°C for 40 s and subsequent exposure to the C_2H_2 carbon precursor at 14 mbar and a flow rate of 200 sccm, at 450°C. An important feature of this process is the low-temperature glass-compatible processing. Figure 2.10b shows a planar view of a honeycomb CNT mesh in low- and high-magnification (upper right inset, Figure 2.10b). The patterns are 300 μm × 300 μm. The CNTs in the vertically aligned mesh structure are highly uniform, dense, and well-aligned perpendicular to the substrate (upper left inset, Figure 2.10b) [24]. Hexagonal arrays were empirically compared with circular and square mesh structures and were found to offer the highest edge field enhancement.

Figure 2.10c illustrates the measurement setup [9]. A phosphor-coated ITO anode was employed during the measurements (as before), shown during electron emission (1 mA.cm^{-2}) in the inset of Figure 2.10d. The *J*–*E* and corresponding FN plot are shown in Figure 2.10d. Typical threshold fields of 1.5 V.μm^{-1} for an emission current density of 1.0 mA.cm^{-2} were observed. A maximum emission current density of 1.5 A.cm^{-2} was recorded at an extraction field of

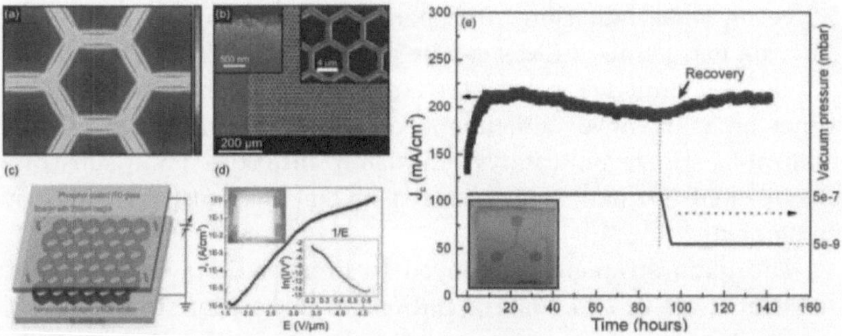

Figure 2.10 Honeycomb emitter. **(a)** Simulated electric field profile. **(b)** SEM micrograph of a fabricated hexagonal mesh. *Upper left inset:* Cross-section SEM micrograph of the CNT forest forming the mesh walls. *Upper right inset:* High-magnification micrograph of 6 hexagonal mesh emitters. **(c)** Measurement set-up. **(d)** *J–E* plot and the corresponding FN plot (*Lower right inset*). *Upper left inset:* Excited phosphor during measurement (1 mA.cm^{-2}). **(e)** Temporal accelerated stability test. *Lower left inset:* Optical micrograph of a fabricated emitter. Chip dimensions = 10 mm × 10 mm [24].

4.8 V.μm^{-1} [24]. Emission recovery was also observed during the stability measurements highlighting the emitters' robustness to CNT degradation and burn-out. The emitter degradation process is thought to be a result of oxygen physisorbtion at high pressures. Oxygen removal increases the emitters' work function leading to a decrease in the emission current. EM inspection of the mesh, post FE, showed no significant degradation in the emitting structure.

Although we have developed a number of high-performance thin-film and periodic-array electron emitters a number of important technological hurdles must still be overcome to achieve successful commercialization. One of these such barriers is the interfacial resistance between the CNTs and the substrate, which results in often unnecessarily large extraction potentials. The use of highly conductive supports obviates this problem.

2.4 CNT Deposition on Conducting Substrates

CNT anchoring onto metallic electrodes, such as metal meshes and etched W tips for example, introduces extremely sharp, high aspect-ratio emitters that enhance the substrates native FE characteristics.

However, weak adhesion often permits CNT de-anchoring during electron extraction. If CNTs can be grown directly on commercially available conductive substrates, resulting in enhanced levels of adhesion, many novel and enhanced FE applications would become realizable. However, catalyst diffusion into the bulk substrate at elevated growth temperatures make this task particularly challenging.

The main strategies employed to deposit CNTs onto metallic substrates are *ex situ* coating and *in situ* deposition. In the *ex situ* approach a variety of binding materials are employed to increase adhesion between the substrate and pre-grown CNTs [35, 49]. These are homogenously dispersed into binders, such as Teflon and Nafion, via damaging acid treatments and ultrasonication, and then cast, sprayed, or dip deposited onto the desired substrates. A huge variety of substrates can be coated in this manner. However, CNT immobilization in the binder medium occludes much of the emitting surface and introduces large interfacial resistances between the CNTs and the metallic support. Uniform CNT alignment is not possible. *In situ* direct deposition negates many of these detrimental features. Here, the catalyst material is traditionally deposited in a powder [25], slurry [42] or vacuum process [50]. Powder and slurry catalyst suspensions require considerable post-processing and, in practice, during the filtration and preparation of the catalyst materials there is major loss of catalytically active species due to attrition. We have successfully deposited dense CNT forests on physical vapor deposited Fe (1 nm) coated Al and Cu foils. However, catalyst deposition using vacuum processes is comparatively costly, roll-to-roll incompatible and can be particularly time consuming.

2.4.1 Stainless Steel Mesh [53]

The catalyst activation of low-cost and widely available, Fe, Ni or Co containing metallic supports offers a possible route to large-scale production of robust, flexible CNT field emitters.

An electrochemical approach was developed to evolve a catalytically active surface on type 304 stainless steel (ss)-mesh (Alfa Aesar, 36% pitch, 50 μm diameter) using a simple potentiostat-type setup and a commercially available metal etchant based on phosphoric acid (H_3PO_4, 66.3 to 70.3 wt.%), acetic acid (CH_3COOH, 8.7 to 10.7 wt.%) and nitric acid (HNO_3, 1.5 to 2.0 wt.%) diluted, and ultrasonically agitated for 10 min in a 1:1 ratio with de-ionized water. CNT synthesis was carried out by PE-CVD at 850°C (5°C.s^{-1}) in an NH_3 (200 sccm) and C_2H_2 (55 sccm) atmosphere (5 mbar) with a low-power, direct current glow discharge plasma (45 W, 70 mA). SEM and high-resolution TEM indicated that the synthesized CNTs, were formed bamboo-type tubes, were 50 μm long, with an aspect ratio >150, and a packing density and self-limited inter-CNT spacing of ~4 × 10^8 cm^{-2} and 0.5 to 5.0 μm, respectively.

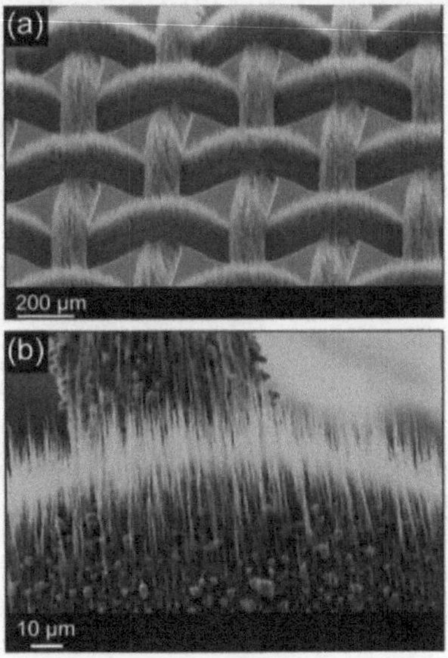

Figure 2.11 **(a)** SEM micrographs of vertically aligned CNTs on electrolyzed stainless steel mesh. **(b)** High-magnification SEM micrograph showing the electrochemically prepared catalyst bed after CNT synthesis.

Native ss-mesh lends itself well to FE applications given its cross-sectional geometry. Indeed, prior to the growth of the CNTs, the ss-mesh exhibited a native field enhancement factor of 1450 (Figure 2.12a). The CNTs increased this to 2900. In addition, the CNT coatings significantly reduced the turn-on fields from 4.5 V.μm^{-1} to approximately 2.5 V.μm^{-1}. Figure 2.12b shows the accelerated lifetime measurement performed at a pressure of 10^{-6} mbar (initial emission current density = 20 mA.cm^{-2}). Stable performance is observed. Modest spikes in the initial emission current are attributed to thermally stimulated degradation caused by gas desorption and the associated lowering of the emitter work function. The interaction of these high currents with the graphitic defects gives rise to substantial heating causing tip burn-out. As shown in the HR-TEM micrographs in the insets of Figure 2.12b, amorphous carbon (*a*-C) coats the CNT tips. *a*-C has greater absorptive capacity than graphitic carbon and we believe it is this *a*-C which gives rise to transient spikes in the lifetime performance. *a*-C graphitization, through annealing treatments has been shown to substantially improve the lifetime of the emitters [5, 27], whilst also permitting higher pressure functionality.

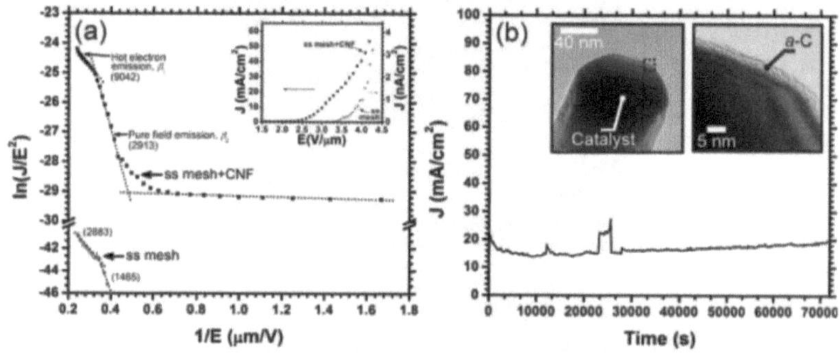

Figure 2.12 FE characteristics of stainless steel mesh coated with vertically aligned carbon nanofibers. **(a)** Comparative FN plot of the as-received ss-mesh and the CNF/ss-mesh composite. *Inset:* The electric-field dependence of the emission current density. **(b)** Temporal stability test conducted at 5×10^{-6} mbar. *Inset:* HR-TEM micrographs of a CNF tip highlighting the local *a*-C, which is believed to degrade the emission sites.

Although these ss-mesh supported CNT emitters offer the benefits of low-cost, simple fabrication and are suitable for a variety of low-end applications, as with the CNT thin films, largely uncontrolled inter-CNT spacing gives rise to undesirable electrostatic shielding effects. In the limiting case, maximum field enhancement can be achieved by the direct deposition, by PE-CVD, of single CNTs onto metallic supports, such as etched commercially available W tips.

2.4.2 W Tips

We have investigated a number of possible approaches to deposit individual CNTs on W tips [13, 28]. Our initial efforts focused on the nanomanipulation of individual W tips and individual CNTs extracted electrostatically from aligned CNT forests. However this process was incredibly time-consuming and required expensive SEM-monitoring and highly skilled operators. Furthermore, the transferred CNTs had poor crystallographic uniformity, irreproducible CNT tip termination and alignment as well as large interfacial resistance associated with a carbon-based glue used to fix the extracted CNT in place [12]. Nonetheless, it was found that these sources performed extremely well with brightness of $\sim 10^9$, energy spreads of 0.2 to 0.5 eV, short-term stabilities of 0.2% at operational temperatures of 700–900°K with noise percentages <0.12% and virtual source sizes of the order of 0.2 nm. Although the emission characteristics of the CNT have been found to be extremely promising the attachment processes were exceedingly cumbersome.

Others [33] have employed thermal CVD to synthesis forests of highly defective CNTs directly on W tips. Nonetheless, for EBL and EM applications it is critical that there is only one CNT that dominates the emission so as to minimize the virtual source size. The first challenge, and arguably the most complex, was to place a single catalyst particle at the tip vertex. The second was to ensure that the CNT is aligned with the W tips emitting axis as the electro-optics in most EM columns can only compensate for beam misalignment within rather narrow tolerances. The third, and final, challenge was to improve the crystallographic quality of the CNTs in an attempt to reduce the axial resistance and interface scattering as well as minimizing the *a*-C content.

We successfully synthesized aligned CNTs, 20 nm in radius and 500 nm in length, directly on wet-etched W tips using PE-CVD, as described elsewhere [28] (Figure 2.13). We found that the deposition reproducibility was the limiting factor, and substantial variations in the number of CNTs deposited on each tip were problematic. COMSOL Multiphysics simulations showed that multiple CNTs at, or near to, the emitter apex gave rise to minimal field augmentation, as long as these secondary CNTs were below the W apex (Figure 2.13a,b). However, as with the periodic CNTs arrays, high packing densities resulted in substantial screening when these secondary CNTs were level with the dominant CNT, and a reduction in the FE performance resulted. In most cases multiple CNTs were deposited. In only 10% of cases was a single CNT found.

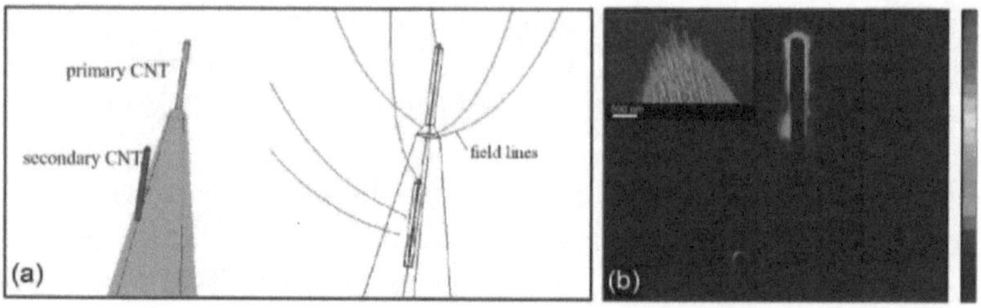

Figure 2.13 **(a)** Schematic showing the presence of a second emitting CNT on a W tip with a dominant emitting CNT at the tips apex. **(b)** Simulated field profile surrounding a W tip coated with two CNT emitters geometrically described in (a) where one CNT is located at the tips apex, the other on the tips side. *Inset:* SEM micrograph of a W tip after PE-CVD coated with multiple CNTs.

2.5 Summary

This review has described some of our on-going work on the development and integration of nanostructure-rich carbon materials, particularly carbon nanotubes and nanofibers, as high-performing alternatives to conventional metallic and Si-based electron emission technologies. By combining our understanding of nanotube synthesis and micro-fabrication we have developed a number of thin film, periodic and ballasted electron sources that have the potential to open up many integrated device possibilities for future FE-based applications including microwave amplifiers, x-ray sources, flexible e-signage, and environmental lighting.

Acknowledgments

The work was partially funded by the European Commission through the AXIS project (Grant No. FP7-SME-1-2007), IST-FET project CANVAD and the NanoICT project. The authors acknowledge funding from KyungHee University International Scholar Scheme, the Scientific Research Foundation, Southeast University, China (Grant No. YBJJ0926), the Schiff Foundation and the Isaac Newton Trust, in addition to the CAPE Associate partners Carl Zeiss, Thales and AIXTRON Nanoinstruments for their collaborative support and the use of their equipment. Support from the WCU program through the

NRF of Korea, funded by MEST (Grant No. R31-2008-000-10029-0) is also gratefully recognized.

References

1. Amaratunga, G. A. J., Baxendale, M. (1999). Field emission from a new form of thin film amorphous carbon having nanoparticle inclusions and carbon nanotubes, *New Diamond Front. Carbon Technol.*, **9**, pp. 31–51.
2. Amaratunga, G. A. J., Silva, S. R. P. (1996). Nitrogen containing hydrogenated amorphous carbon for thin-film field emission cathodes, *Appl. Phys. Lett.*, **68**, pp. 2529–2531.
3. Baker, F. S., Williams, J. (1972). Field-emission from carbon fibers — new electron source, *Nature*, **239**, pp. 96–97.
4. Bandis, C., Pate, B. B. (1995). Photoelectric emission from negative-electron-affinity diamond (111) surfaces: Exciton breakup versus conduction-band emission, *Phy. Rev. B*, **52**, p 12056.
5. Baoqing, Z., Guangyong, X. (2006). Enhancement of field emission of aligned carbon nanotubes by thermal oxidation, *Appl. Phys. Lett.*, **89**, pp. 223119-1-223119-223119-3.
6. Bingbing, L., Sundqvist, B. (2001). Electric resistance of single-walled carbon nanotubes under hydrostatic pressure, *Solid State Commun.*, **118**, pp. 31–36.
7. Bonard, J. M., Stockli, T. (2001). Field emission from cylindrical carbon nanotube cathodes: possibilities for luminescent tubes, *Appl. Phys. Lett.*, **78**, pp. 2775–2777.
8. Brodie, I., Spindt, C. A. (1992). *Vacuum Microelectronics, Advances in Electronics and Electron Physics*, **83**, pp. 1–106.
9. Chi, L., Yan, Z. (2010). High emission current density, vertically aligned carbon nanotube mesh, field emitter array, *Appl. Phys. Lett.*, **97**, pp. 113107-3.
10. Choi, W. B., Jin, Y. W. (2001). Electrophoresis deposition of carbon nanotubes for triode-type field emission display, *Appl. Phys. Lett.*, **78**, pp. 1547–1549.
11. Collins, P. G. and Zettl, A. (1997). Unique characteristics of cold cathode carbon-nanotube-matrix field emitters, *Phys. Rev. B*, **55**, p 9391.
12. de Jonge, N., Allioux, M. (2004). Characterization of the field emission properties of individual thin carbon nanotubes, *Appl. Phys. Lett.*, **85**, pp. 1607–1609.

13. de Jonge, N., Lamy, Y. (2003). High brightness electron beam from a multi-walled carbon nanotube (vol 420, pg 393, 2002), *Nature*, **423**, pp. 461–461.
14. Dean, K. A. and Chalamala, B. R. (1999). The environmental stability of field emission from single-walled carbon nanotubes, *Appl. Phys. Lett.*, **75**, pp. 3017–3019.
15. Djubua, B. C. and Chubun, N. N. (1991). Emission properties of spindt-type cold cathodes with different emission cone material, *IEEE Trans. Electron Devices*, **38**, pp. 2314–2316.
16. Dresselhaus, M. S. and Avouris, P. (2001). Introduction to carbon materials research, *Carbon Nanotubes*, **80**, pp. 1–9.
17. Fowler, R. H. and Nordheim, L. (1928). Electron emission in intense electric fields, *Proceedings of the Royal Society of London Series A-Containing Papers of a Mathematical and Physical Character*, **119**, pp. 173–181.
18. Geis, M. W., Efremow, N. N. (1991). Diamond cold-cathode, *IEEE Electron Device Lett.*, **12**, pp. 456–459.
19. Gerber, J., Robertson, J. (1996). Role of surface diffusion processes during bias-enhanced nucleation of diamond on Si, *Diamond Relat. Mater.*, **5**, pp. 261–265.
20. Gilkes, K. W. R., Sands, H. S. (1998). Direct observation of sp(3) bonding in tetrahedral amorphous carbon UV raman spectroscopy, *J. Non-Cryst. Solids*, **227**, pp. 612–616.
21. Groning, O., Kuttel, O. M. (1999). Field emission properties of nanocrystalline chemically vapor deposited-diamond films, *J. Vacuum Sci. Technol. B*, **17**, pp. 1970–1986.
22. Himpsel, F. J., Knapp, J. A. (1979). Quantum photoyield of diamond(111) — a stable negative-affinity emitter, *Phys. Rev. B*, **20**, p 624.
23. Jung, S. M., Hahn, J. (2006). Clean carbon nanotube field emitters aligned horizontally, *Nano Lett.*, **6**, pp. 1569–1573.
24. Li, C., Zhang, Y. (2010). Stable, self-ballasting field emission from zinc oxide nanowires grown on an array of vertically aligned carbon nanofibers, *Appl. Phys. Lett.*, **96**, pp. 143114-3.
25. Liu, Y., Qian, W. Z. (2008). The confined growth of double-walled carbon nanotubes in porous catalysts by chemical vapor deposition, *Carbon*, **46**, pp. 1860–1868.
26. Mann, M., Teo, K. B. K. (2008). Carbon nanotubes as electron sources, *Materials Science Foundations*. http://www-g.eng.cam.ac.uk/cnt/oldsite/papers/mark_review.pdf. Internal publication.

27. Mann, M., Teo, K. B. K. (2008). Stabilization of carbon nanotube field emitters, *Proceedings of the Institution of Mechanical Engineers, Part N (J. Nanoeng. Nanosystems)*, **222**, pp. 95–99.

28. Mann, M., Teo, K. B. K. (2006). Direct growth of multi-walled carbon nanotubes on sharp tips for electron microscopy, *Nano*, **1**, pp. 35–39.

29. Milne, W. I., Teo, K. B. K. (2004). Carbon nanotubes as field emission sources, *J. Mater. Chem.*, **14**, pp. 933–943.

30. Milne, W. I., Teo, K. B. K. (2006). Carbon nanotubes as electron sources, *Phys. Status Solidi A Appl. Mater. Sci.*, **203**, pp. 1058–1063.

31. Musa, I., Munindrasdasa, D. A. I. (1998). Ultra-low-threshold field emission from conjugated polymers, *Nature*, **395**, pp. 362–365.

32. Nation, J. A., Schachter, L. (1999). Advances in cold cathode physics and technology, *Proc. IEEE*, **87**, pp. 865–889.

33. Riley, D. J., Mann, M. (2003). Helium detection via field ionization from carbon nanotubes, *Nano Lett.*, **3**, pp. 1455–1458.

34. Robertson, J. (1999). Mechanisms of electron field emission from diamond, diamond-like carbon, and nanostructured carbon, *J. Vacuum Sci. Technol. B*, **17**, pp. 659–665.

35. Rubianes, M. D. and Rivas, G. A. (2003). Carbon nanotubes paste electrode, *Electrochem. Commun.*, **5**, pp. 689–694.

36. Saito, Y., Hamaguchi, K. (1998). Field emission from multi-walled carbon nanotubes and its application to electron tubes, *Appl. Phys. A Mater. Sci. Process.*, **67**, pp. 95–100.

37. Satyanarayana, B. S., Hart, A. (1997). Field emission from tetrahedral amorphous carbon, *Appl. Phys. Lett.*, **71**, pp. 1430–1432.

38. Semet, V., Binh, V. T. (2002). Field electron emission from individual carbon nanotubes of a vertically aligned array, *Appl. Phys. Lett.*, **81**, pp. 343–345.

39. Shim, J. Y., Baik, H. K. (2000). Effect of nondiamond carbon on electron transport path of field-emitted electrons from undoped polycrystalline diamond films, *J. Vacuum Sci. Technol. A*, **18**, pp. 1977–1982.

40. Smith, R. C., Carey, J. D. (2005). Charge transport effects in field emission from carbon nanotube–polymer composites, *Appl. Phys. Lett.*, **87**, pp. 263105-3.

41. Talin, A. A., Felter, T. E. (1996). Electron field emission from amorphous tetrahedrally bonded carbon films, *J. Vacuum Sci. Technol. A Vacuum Surf. Films*, **14**, pp. 1719–1722.

42. Tang, S. H., Sun, G. Q. (2010). Double-walled carbon nanotubes as catalyst support in direct methanol fuel cells, *J. Electrochem. Soc.*, **157**, pp. B1321–B1325.

43. Teo, K. B. K., Chhowalla, M. (2001). Uniform patterned growth of carbon nanotubes without surface carbon, *Appl. Phys. Lett.*, **79**, pp. 1534–1536.

44. Teo, K. B. K., Minoux, E. (2005). Microwave devices — carbon nanotubes as cold cathodes, *Nature*, **437**, pp. 968–968.

45. Teo, K. B. K., Rodil, S. E. (2001). Effect of graphitic inclusions on the optical gap of tetrahedral amorphous carbon films, *J. Appl. Phys.*, **89**, pp. 3706–3710.

46. Unalan, H. E., Hiralal, P. (2008). Rapid synthesis of aligned zinc oxide nanowires, *Nanotechnology*, **19**, p 255608.

47. van der Weide, J., Zhang, Z. (1994). Negative-electron-affinity effects on the diamond (100) surface, *Phy. Rev. B*, **50**, p 5803.

48. Wang, C., Garcia, A. (1991). Cold field-emission from CVD diamond films observed in emission electron-microscopy, *Electron. Lett*, **27**, pp. 1459–1461.

49. Wang, J., Musameh, M. (2003). Solubilization of carbon nanotubes by nafion toward the preparation of amperometric biosensors, *J. Am. Chem. Soc.*, **125**, pp. 2408–2409.

50. Zhang, C., Pisana, S. (2008). Growth of aligned millimeter-long carbon nanotube by chemical vapor deposition, *Diamond Related Mater.*, **17**, pp. 1447–1451.

51. Zhu, W., Kochanski, G. P. (1995). Defect-enhanced electron field-emission from chemical-vapor-deposited diamond, *J. Appl. Phys.*, **78**, pp. 2707–2711.

52. Li, C., Zhang, Y., Cole, M.T., Shivareddy, S.G., Barnard, J.S., Lei, W., Wang, B., Pribat, D., Amaratunga, G. A. J. , and Milne, W. I. (2012). Hot electron field emission via individually transistor-ballasted carbon nanotube arrays, *ACS Nano*. DOI: 10.1021/nn300111t, in press.

53. Cole, M.T., Hou, K., Warner, J. H., Barnard, J. S., Ying, K., Zhang, Y., Li, C., Teo, K. B. K., Milne, W.I. (2011). *In-situ* deposition of sparse vertically aligned carbon nanofibres on catalytically activated stainless steel mesh for field emission applications. *Diamond Related Mater.*, **23**, pp. 66–71.

Chapter 3

Plasma-Enhanced Chemical Vapor Deposition of Graphene Nanostructures

Timothy van der Laan,[a,b] Shailesh Kumar,[a] and Kostya (Ken) Ostrikov[a,b,*]

[a] *Plasma Nanoscience Centre Australia (PNCA), CSIRO Material Science and Engineering, P.O. Box 218, Lindfield NSW 2070, Australia*

[b] *Plasma Nanoscience, Complex Systems, School of Physics, The University of Sydney NSW 2006, Australia*

*Kostya.Ostrikov@csiro.au

3.1 Introduction

Graphene has received great interest from researchers all over the world owing to its unique properties. Much of the excitement surrounding graphene is due to its remarkable properties and inherent quantum effects. These effects and properties make it a desirable material for the fabrication of new devices. Graphene has a plethora of potential uses including gas and molecular sensors, electronics, spintronics and optics [1–7]. Interestingly, some of these properties have been known about since before the material was even isolated due to a considerable amount of theoretical work and

Advances in Nanodevices and Nanofabrication: Selected Publications from Symposium of Nanodevices and Nanofabrication in ICMAT2011
Edited by Qing Zhang, W. I. Milne, and Jianping Zou

ISBN 978-981-4364-54-6 (Hardcover), 978-981-4364-55-3 (eBook)
www.panstanford.com

simulations. The material was to some extent a condensed matter modelers "toy" as it was used as a benchmark 2D material. Graphene had been used for a long time as the fundamental building block of many other carbon structures.

The isolation of graphene had been deemed impossible as it was considered that the structure would be thermodynamically unstable [8]. This was further emphasised by the discovery of more thermodynamically stable structures including fullerene and nanotubes. These curved graphene-based structures led researchers to believe that it would not be possible for graphene to hold its shape if it was isolated and it would undergo spontaneous curving. This theory was, however, disproven and it was through very tedious trial and error mechanical exfoliation experiments that graphene was eventually isolated and its properties verified experimentally [9]. Despite its isolation and the confirmation of its incredible properties graphene still remains largely in the lab due to the great difficulty in producing high-quality and large-area films.

Implementing graphene in new devices is going to require an effective approach to synthesizing graphene. This chapter presents plasma-enhanced chemical vapor deposition (PECVD) as a promising approach. To this end, the chapter is divided into three sections: (1) a summary of other methods of synthesis, (2) a discussion of why plasmas may be a useful tool, and (3) a summary of the current status of plasma-based synthesis of graphene, including recent original results produced at the Plasma Nanoscience Centre Australia (PNCA).

3.2 Successful Approaches to Synthesizing Graphene

There have been a number of successful approaches to synthesizing graphene. They can be broken down into three different tactics, mechanical exfoliation, chemical synthesis and supported growth. There is still, however, a large problem facing all synthetic methods, the difficulty of confirming the nanostructure produced is in fact graphene. Currently the only way to first determine that a grown film is indeed graphene and to then characterize it is through a combination of experiments. No one single test can determine presence of and then thickness, size and purity of graphene. The most commonly used characterization techniques include optical microscopy, Raman spectroscopy and transmission electron microscopy (TEM) or atomic force microscopy (AFM).

Drawbacks exist, however, in both the required number and nature of these techniques. For instance, the first two techniques require the use of a very specific substrate, Si with a native oxide coating. In general, a 300 nm native oxide layer is used; however, many other thicknesses can be used depending on the wavelength of light being used [10]. Raman spectroscopy is only fully characterized on a SiO_2 substrate (although a number of reports are now available on the interaction of graphene with other substrates and the subsequent changes to the spectra). TEM techniques also require specific frameworks which allow the desired interaction with electron beams. AFM has restraints in that it requires a smooth surface to be able to differentiate between the bare surface and graphene covered surface. Generally the only way to ensure this is by conducting the characterization on a newly smoothed substrate and not the one used during synthesis. This means if one were to conduct the above characterizations, transfer to a SiO_2/Si substrate is required as well as to a TEM grid or an atomically smooth substrate. Importantly, no matter what techniques are used, it is more than likely a transfer will be required at some stage. This then leads to the question whether the graphene was indeed synthesized on the surface during treatment or whether through subsequent transferring processes few-layer graphene (FLG) or graphite was broken down into graphene sheets. This is further emphasized by single-layer graphene generally being mixed with a mess of multi layer sheets, which are considered impurities [11]. This problem has no simple solution nor is one presented here; it is, however, noted as this is a problem that exists in the synthesis of graphene that also holds back commercial applications of graphene.

3.2.1 Mechanical Exfoliation

Mechanical exfoliation (ME) is an approach to fabricating graphene where layers of graphite are mechanically separated until a single monolayer (graphene) is left. This technique is effective as it makes use of the fact that the only force holding two sheets of carbon together are weak van der Waals forces. The cumulative effect of these forces is that a force of the order of 300 $nN.\mu m^{-2}$ [12] is required to separate one layer from another. This approach was demonstrated in the way graphene was first isolated [9]. Through the use of common adhesive tape and repetitive stick and peel processes a 1 μm thick graphite flake was reduced to a single monolayer [9].

A drawback of this ME technique arises due to residues left from the glue on the tape and therefore subsequent treatment (usually heating) is required to remove this contamination. The above synthesis is the most common approach to ME. However, there are other approaches. One of these other approaches uses the application of a high voltage to the graphite. This causes adhesion of graphene to a substrate and subsequently when the graphite is removed graphene remains stuck to the substrate [13]. This is an improvement over other ME techniques as it removes glue contamination, as none is used.

Despite the apparent archaic nature of this method of synthesizing graphene it does produce the very high-quality graphene films [14]. This is because the quality of the graphene is intimately linked to the quality of the initial piece of graphite and single crystalline graphite sources are quite readily available. A marked advantage of this particular approach to isolating graphene is that the graphene can be "stuck" directly to the surface required for analysis so no further transfer is required. Obviously the major drawback of this technique is the required manpower and the great variation in whether or not graphene is isolated or not. In general, when graphene is isolated, it is mixed with multi-layered graphene and can be difficult to find [11].

3.2.2 Chemical Synthesis

Chemical synthesis involves the reduction of graphene oxide into graphene. While first synthesized in 1859 by Brodie using volatile reagents, graphene oxide is now easily produced with a safe and simple chemical reaction established in 1958 by Hummers [15]. This process simply involves combining a graphite dispersion with concentrated sulphuric acid, sodium nitrate and potassium permanganate and heating to 45°C for a couple of hours. Graphene oxide is simply the product of oxidation of graphite and contains varying amounts of carbon, oxygen and hydrogen depending on the efficiency of the synthesis. Once graphene oxide sheets are isolated, they then require reduction to convert them to graphene.

A variety of approaches to reduce graphene oxide have yet to be successful in producing 100% graphene films. These approaches range from highly intricate, multi-step reduction processes to simple deoxygenation in alkaline solutions [16, 17]. While these approaches do not produce pure graphene films, considerable portions of the samples are reduced [18] and, as such, these are techniques being pursued by researchers as even these partial graphene films have quite low resistances and high transparency.

3.2.3 Supported Growth

Supported growth is the most widely researched process for fabricating graphene. This is because supported growth processes present a method that can be readily scaled up. There are two primary avenues to producing graphene in a supported growth process: (1) thermally decomposing carbides and (2) chemical vapor deposition (CVD) on metals or carbides from carbon containing precursor gases.

Thermal decomposition of carbides requires very high temperatures, typically higher than 1000°C and has been successful with a number of different carbides. Using temperatures of 1000°C and n-type doped 6H-SiC(0001), graphene has been grown [19]. Graphene coverage of the surface was incomplete; however, the graphene islands formed are, for the most part, high-quality graphene. This is a common synthesis of graphene from SiC with similar results being produced at higher temperatures in a number of studies [20, 21]. There have also been a few reports of the growth of graphene on TiC(111) [22], TiC(410) [22] and TaC(111)[23]. However, each of these reported studies has required a particular crystalline growth face and occurs at temperatures much greater than 1000°C.

CVD growth differs from thermal decomposition in that a source of carbon is required as there is no carbon contained in the substrate used. Common carbon sources include hydrocarbons and fluorocarbons injected in the gas phase. There a wide variety of different catalysts that have been used to grow graphene including Pt(111) at 1000 K [24], Ru(001) at 1270 K [25], Ni(111) at 1230 K [26] , Ir(111) at 970 K [27], Rh(111) at 920 K [28] and Co(0001) at 600 K [29]. Most growth occurs on a single crystalline face of a metal substrate. There are some reports of growth on polycrystalline Ni [30], Pt [31] Co [32] and Cu [33]. The proposed growth mechanism for most CVD processes relies on the dissolution of gas phase carbon precursors into the metal and subsequent precipitation as the catalyst is rapidly cooled. The problem with this growth process is that it can be hard to control the number of graphene layers that are expelled on precipitation. It should be noted that this is not the only growth mechanism. An example is copper; here graphene is grown in a surface-catalysed process whereby the carbon interacts on the surface [33]. This is important as it means that it is a self-limiting process that once graphene fully covers the surface there is no more catalyst available to increase the number of layers.

There have been no reports of growth on non-crystalline metal catalysts. All reports either involve an annealing step or begin with crystalline films. It should also be noted that the morphology of the grown films is constrained to that of the catalyst being used and as such cannot be controlled to any large extent. This is relevant for two reasons. First, it means that the particular surface energy of the catalyst plays some role in the formation of graphene, which is confirmed by the variety of temperatures at which the different catalysts function. Additionally, it also means that the growth of graphene in such a process is either limited to expensive single-crystal materials or that high temperature annealing is required to produce a polycrystalline phase for growth. These processes, which in the long run are required for large-scale implementation by industry, have two major flaws: the very high temperatures required and the cost of catalysts. Ideally, to overcome these flaws in CVD, low-temperature synthesis and economically viable substrates need to be developed.

3.2.4 Summary of Graphene Synthesis

The desire to upscale these processes and incorporate graphene into devices is an area of ongoing research. As outlined in Table 3.1 currently there is yet no ideal technique to fabricate graphene suitable for commercial device applications. To be appropriate for device integration, a technique is required that is controllable, cheap and easy to upscale. None of the current techniques meets these requirements. The most promising, however, is supported growth. If this technique can be optimized this approach would meet the requirements for industrial scale growth of graphene. One, largely unexplored method to optimize supported growth of graphene is

PECVD. This technique may be useful as it has the potential to meet all the requirements that have been outlined for industrial scale device integration due to the great advantages plasma-enhanced processes have over standard neutral gas CVD.

Table 3.1 Summary of different approaches to synthesizing graphene

Technique	Advantage	Disadvantage
Mechanical Exfoliation	Cheap and easy	Highly variational
	Simple Equipment	Uneven films
		Labor Intensive
Chemical Synthesis	Easy up-scaling	Fragile and volatile reagents
	Quick process	Only partially reacted products
Supported Growth	Even films	Hard to control morphology
	Large areas	High Temperatures

3.3 Advantages of Plasma-Enhanced Chemical Vapor Deposition Approaches

3.3.1 Plasma-Related Effects

The application of a plasma to enhance the growth of nanoscale objects is an area which is not only being researched extensively but also producing exciting results. Most important, the major thrust of the research is centered on the advantages that are gained through the addition of a plasma to nanoscience, i.e., why would one want to use plasma-based nanofabrication techniques. The answer, plasmas add a number of interesting properties to a system that can be harnessed leading to the formation of "better" nanostructures due to the way energy and matter can be manipulated [34–36]. The unique properties of a plasma are that it allows enhanced control of the delivery of matter and energy to a surface. The implications and attempts to control these properties are topics of ongoing research.

3.3.1.1 Building units

The greater control of matter in a plasma is primarily evident in the much greater number of building units (BUs) that can be utilized to form structures in a plasma system. One moves from growing structures using predominantly neutral molecular states to a plethora of different BUs including atoms, ions, radicals, agglomerates, the original molecule and also a number of other molecules. This means that when constructing a nanostructure in a plasma, researchers are no longer confined to using only one simple BU. This makes the construction of more exotic and more stable nanostructures possible [37]. A useful analogy would be to compare the structures you can make with one type of Lego block to the structures you can make with a dozen different types. The greater number of different BUs constructed in the plasma can be selected and brought to the surface by modifying the plasma parameters. This means that undesirable species that are produced in the plasma that may lead to the growth of unwanted structures can be isolated from the surface. By doing so, the impact of undesirable BUs has a minimal impact on the growth of nanostructures and desirable BUs can be selectively targeted. Control over the BUs, however, is not the only advantage presented by implementing a plasma to nanoscale synthesis.

3.3.1.2 Plasma heating

Plasma heating is a beneficial by product of implementing a plasma in a growth process. Quite a simple concept, plasma heating is a consequence of the very nature of a plasma. That is, they are a collection of charged and non-charged particles with differing energies. When a plasma is generated a lot of energy is transferred to the (in general) gas phase. Not only is this energy injection responsible for the breakdown of molecular species into the wide variety of BUs in a plasma but it also provides kinetic energy to the constituent species, most of this to the charged species. In the case of low-temperature plasmas (commonly used in PECVD) a large fraction of the energy of the plasma is intrinsic to the ions. This is a result of the much lighter and mobile electrons in the plasma being readily lost to the environment by colliding with the chamber or lost to the vacuum. The ions on the other hand, being much heavier and less mobile, are not so easily lost. This means that when a bias is applied to the substrate and the ions are attracted to the surface a significant proportion of the energy used to produce the plasma is transferred to the surface. Thus through the bombardment of the surface a large amount of kinetic energy is transferred from the plasma to the substrate. The result of this kinetic energy transfer is what is meant by plasma heating whereby the temperature of the material exposed to the plasma is increased [38]. This means that the plasma can be used as an effective source of heating and, if controlled correctly, may even remove the need for an external heating source (such as the large furnaces common to CVD processes) altogether.

This plasma heating could be of great importance to the growth of graphene in that this heating is localized to the upper layers of the substrate. This means that more of the energy applied to the system is focused on the surface where the graphene is growing. This reduces the wasted heat in CVD systems where the entire chamber is heated not just the area where heat is required. Not only this but as it is only the first few layers of the substrate that are heated the dissolution of carbon species into the material is limited which would be beneficial if graphene growth occurs through a precipitation method. If the predominant route of growth of graphene is a surface diffusive process then this too is beneficial as then energy is not wasted in heating the entire substrate. Generally the heating provided by a plasma is of great benefit in minimizing the changes to the bulk material due to heating and lowering the activation temperature required for synthesis.

3.3.1.3 Surface activation

Surface activation is another useful property of the plasma that can enhance the growth of nanofeatures on a surface. Surface activation is a secondary property of bombardment by charged species. It is a process by which an inert surface is "activated" by the reconfiguration of bonding on its surface. When a charged particle impacts the surface, there are a number of possible scenarios that may take place, including reflection and loss of kinetic energy; sticking and incorporation into growing features; surface reaction with substrate which may produce dangling or weaker bonds; and ablation. Each of these plays some role in activating the surface as increasing the number of dangling bonds, unfavorable reaction and defects all serve to increase the surface energy of the exposed surface [39]. This increased surface energy means that whilst in the neutral case, long-range movements of adatoms and nucleation of structures would be considered unlikely, they become more pronounced in a plasma [40]. Thus the increased surface energy results in the favorable growth of structures on the surface.

Surface activation in a plasma may play a pivotal role in stabilizing the growing graphene films. It is widely accepted that graphene is not the most stable configuration of carbon atoms and precautions have to be taken in the growth of graphene to prevent the atoms from reconfiguring. Generally this is achieved by controlling the thermodynamics and kinetics of the growth. This is seen in the way graphene is grown in a standard CVD process whereby the system is elevated to a high temperature where graphene growth becomes possible. By quickly drawing the heat away from the system via rapid cooling atoms quickly lose their intrinsic energy and thus no longer have the required energy to initiate reconfiguration of the graphene film. In the case of surface activation through a plasma it may be possible to make the growing surface unstable enough that the coverage of this surface with a graphene film would actually lower the surfaces overall energy. It may also be possible that diffusion of carbon atoms from the catalyst/film interface onto the film or into the substrate could be prevented as activating the surface enhances the "stickiness" of the catalyst's surface.

3.3.1.4 Usefulness of plasmas

Owing to the specific properties of plasmas in the context of nanofabrication, a large number of studies have sought to verify that the effects outlined above do enhance growth of nanostructures [41–46]. The investigations carried out both theoretically and experimentally have confirmed on numerous occasions and in a variety of different contexts just how powerful plasma-based techniques can be. Successful applications range from the production of quantum dots, nanowires, nanotubes to thin films, all of which have had their fabrication enhanced in some way by plasmas. Thus the benefits of plasma properties to enhancing the growth of graphene would appear to be a logical avenue of investigation. The inductively coupled plasma set up of the PNCA run at RF frequency of 13.56 MHz is shown schematically in Figure 3.1 and in operation in Figure 3.2. These figures illustrate the ease in which one can add a plasma to enhance processes. The schematic shown for an inductively coupled plasma can be easily applied to many existing chambers. This is simply done by wrapping the coil around the tube (common to most CVD furnaces) near to the gas inlet and using a suitable material for the tube, e.g. quartz.

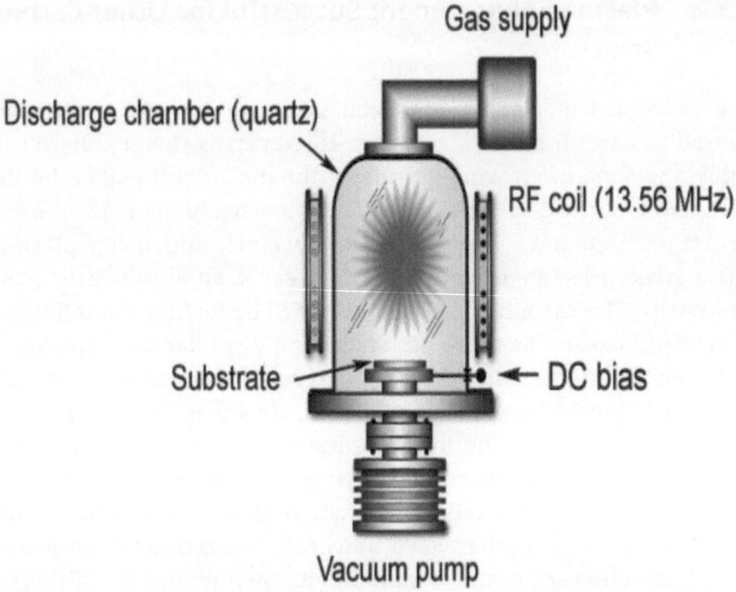

Figure 3.1 Schematic of plasma setup.

Figure 3.2 Plasma in operation.

3.3.2 Plasma Enhancement Successful for Other Carbon Structures

The successes of plasma-enhanced processes are not specifically limited to carbon-based structures. However, as the goal here is to apply a plasma to growing graphene, the discussion will be limited to carbon nanostructures. There are a variety of carbon-based structures that have been successfully fabricated using plasmas. These structures range from diamond to carbon nanotubes to carbon nanowalls. The carbon nanostructures will be further limited to sp^2-hybridized carbon as these are structurally similar to graphene. In fact, one could consider these structures to be constructed from building blocks of graphene. It should, however, be noted that in the case of diamond, the implementation of plasmas has lowered the required temperature and pressure for synthesis. This has been demonstrated by the successful growth of diamond on a plastic film at 100°C by using a microwave powered, low-temperature plasma [47]. Each different example requires its own unique conditions to grow, which is another benefit of using plasma-enhanced fabrication methods as conditions can be altered to the particular requirements of a desired structure.

3.3.2.1 Carbon nanotubes

Carbon nanotubes are the salient example of where a carbon structure's growth can benefit from plasma enhancement. This is because nanotubes are not only a material which has been extensively researched with respect to plasma-enhanced nanofabrication, but they are also one of the hottest topics for research in physics in general due to their wide variety of applications. Work on nanotubes is also not confined to just the laboratory with considerable work being conducted with simulations. The application of a plasma with an electric field has been shown to be successful in improving the vertical alignment of the tubes compared to the curled tubes grown via standard CVD [48]. This has been explained to some extent through simulations which demonstrate that the growth of the tubes in a plasma includes additional carbon addition on the tube and that the growth need not rely as heavily on surface diffusion to source carbon for the growing structure. In addition, it is noted that the growth of the nanotubes can occur at low temperatures, at a faster rate and the tubes can grow longer by utilizing a plasma [49]. Experimentally it has been shown that temperatures as low as a couple of hundred degrees centigrade can be used to deposit carbon nanotubes when a plasma is used compared to temperatures around 700–800°C in a neutral system [50]. The predominant factor that allows the enhanced growth of carbon nanotubes in a plasma environment is that the BUs have been modified and supplied with sufficient energy [51]. In the plasma system, the hydrocarbons used to grow carbon nanotubes now exist as radicals, ions and other species. The standard model for carbon nanotubes growth presented by Baker *et al.* [52] describes the activation of nanotubes being reliant on the decomposition of hydrocarbons at the catalyst surface. So in a plasma when the hydrocarbons are already partially or completely decomposed, the advantages of using a plasma are obvious [53].

One may consider the analogy with the growth of graphene whereby precursor gases, most commonly hydrocarbons, must be broken down. This is the case no matter the growth mechanism as not only will the final product contain no hydrogen but also the dissolution and diffusion energy barriers are smaller for atomic carbon than for hydrocarbons. Thus by using a plasma, the step where by the hydrocarbon must be broken down on the growing surface is eliminated. Further the diffusion of carbon atoms is enhanced allowing better carbon delivery. The faster growth of carbon nanotubes may also mean that growth can be more kinetically driven, which will allow better graphene growth as for graphene growth to occur kinetics must be greatly controlled.

3.3.2.2 Vertically aligned graphene nanosheets and carbon nanowalls

The advantages of a plasma have also extended to the growth of carbon nanowalls and vertically aligned graphene nanosheets. This advance is important as this structure is in effect a vertically aligned interwoven graphene network. These networks are found to grow on a variety of substrates and most interestingly only in plasma environments [54]. It is postulated that the reason why these structures only grow in a plasma environment is a result of the strong vertical alignment of nanostructures when a bias is applied to the substrate [55] though the actual growth mechanism is unknown. Simulations again demonstrate that the way a plasma can deposit material directly onto the growing structure is likely playing a major role in the process as simple diffusion over the surface cannot provide enough material to the structures to continue growth [56]. Interestingly, the role of the catalyst seems to play a lesser role in the growth of the structure compared to other carbon nanostructures. This is demonstrated by growth of the structures on catalyst-free surfaces [57] suggesting that the growth is likely a result of the strong interaction of the plasma with the growing surface.

Work conducted in our group (to be published) has also demonstrated this further by minimizing the temperature required for growth. Carbon nanowalls can be grown without the need of any external heating source and without a catalyst. The morphology and subsequent properties of these structures can then be altered by altering plasma parameters. This culminates all the benefits of a plasma into a single example as it shows that a plasma, on its own, can be used to grow and control the growth of a graphene like structure. One problem that still currently exists is the thickness of these walls is limited to a few layers of graphene. Current explanations for this lean toward bi- or single-layered walls being too stressed to remain vertically aligned and thus collapsing. These explanations find merit as even horizontal graphene requires a surface to stabilize it and prevent it curving and forming other materials.

Furthermore, it is generally noted that the growth of sp^2 carbon structures in a plasma environment is enhanced. This is a consequence of the preferential etching of sp^3 carbon from sp^2 carbon in amorphous carbon by atomic hydrogen [58]. The great amount of atomic hydrogen produced in PECVD systems thus results in fewer sp^3 carbon impurities when growing sp^2-hybridized carbon structures. The growth of structures such as these provides strong support for the growth of graphene in a plasma environment.

3.4 Current Status of PECVD of Graphene

3.4.1 General Review

There are currently very few reports of growth of graphene in a plasma environment making this a greatly unexplored field of research. Reports of graphene growth are contained to only a few metal catalysts and no success on catalyst free surfaces. Graphene growth has, however, been reported for both microwave and radio frequency plasmas. Evidence presented in many of these reports, however, is not very convincing with analysis done post transfer and also much of the evidence being consistent with few-layer graphene growth rather than single-layer graphene. There are also a great amount of reports of few-layer graphene flakes though it is hard to discern whether these are horizontal or vertically aligned structures as analysis of the structures grown is almost exclusively done post transferring of the grown structures. Here we will limit the discussion to only those reports which claim to have grown graphene.

Nickel due to the great many publications on the material appears to be the most popular catalyst for CVD growth of graphene. Nickel has been shown to be a useful catalyst in standard CVD so its extension to plasma CVD seemed logical. Successful growth has been demonstrated on both polycrystalline and single crystalline surfaces with comparable growth temperatures, pressures and plasma powers. Growth time is very short compared to standard CVD growth and rapid cooling is not required and growth occurs at a lower temperature of 650–700°C [59]. It should be noted that this produced a lot of few-layer graphene as well as single-layer graphene; the average film thickness was estimated at 7 layers. Figure 3.3 shows two Raman spectra presented as evidence in this report. It is to be noted that the Raman spectrum of graphene is shown in Figure 3.3b only after transfer of the material to a clean substrate. This highlights the issues in how characterization is carried out, is it graphene or few-layer graphene grown with graphene only evident after transfer of the grown film for analysis?

This report whilst providing proof of concept, a plasma can be implemented to grow graphene, contains very little information about the effects the plasma has on the growth. The only note made is that there is a 300°C difference in growth temperature by utilizing a remote 13.54 MHz plasma at 250 W and keeping all other parameters constant [59]. Further reports of Nickel growth using PECVD provide little more information except to provide varying experimental condition where growth is possible. High-quality large area growth of graphene on a Nickel catalyst has also been reported [60], however, the Raman spectra provided as evidence is not convincing. Again this report has little to say about the manner in which a plasma enhances the growth process.

Many reports whilst claiming to have grown graphene seem to provide evidence to the contrary and in fact provide more evidence for the growth of few-layer graphene. A clear distinction must be made here as graphene composing of a single layer of carbon has a host of properties that are not found in few-layer graphene. Whilst few-layer graphene may be useful in a great number of applications for PECVD to become a useful technique in synthesizing graphene single-layer graphene is required and few-layer graphene impurities kept to a minimum. The most important step in achieving this will be a clear description of the growth process in a PECVD system. Once this is achieved then procedures can be optimized for growing single-layer graphene.

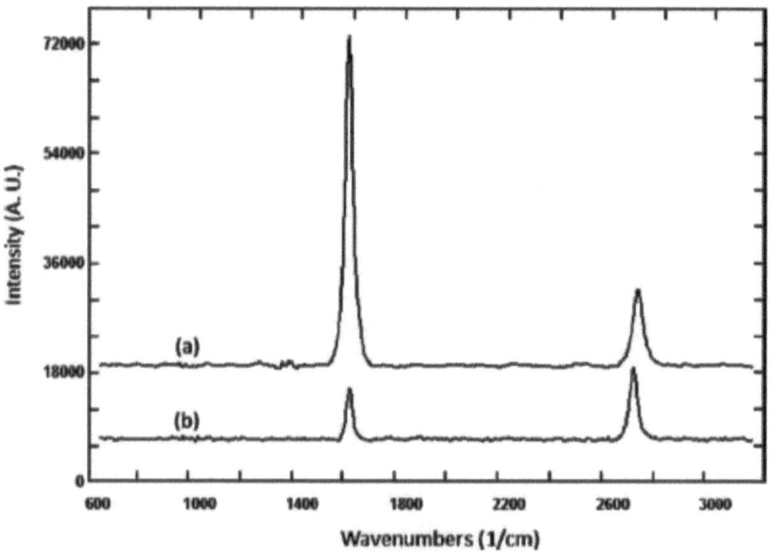

Figure 3.3 (a) Raman spectrum of film grown on Ni (b) Raman spectrum of film after transfer to clean SiO_2/Si substrate. Spectrum (a) indicative of few-layer graphene (b) indicative of graphene [59]. Reproduced with permission from American Institute of Physics.

The great number of reports of high-quality few-layer graphene should, however, be noted as it is these reports that lend weight to our argument that PECVD can be a powerful tool in growing graphene. It has been shown that these films can be grown on Co [61], Al and Cu [62] and Ni [63]. The growth temperature for these lower than that of CVD processes so one immediate success noted in implementing a plasma is that growth temperatures are reduced. This is important as the major waste of energy in a CVD process is in the heating of the system. Each of the mentioned reports also exhibits areas with some graphene. Despite the film containing few-layer graphene, this shows that in principle, with an optimized process, these unwanted layers could be reduced. Growth using non-crystalline surfaces is also promising especially as this means cheap substrates are possible. Not only this but also it minimizes the need to anneal for a period of time at a higher temperature, which further reduces the demand on heating in the overall process of growing graphene.

3.4.2 Authors' Recent Contributions

Catalyst-free growth of few-layer graphene films has been reported [64]; there is, however, no conclusive evidence for graphene's presence. Work to be published by the authors has shown graphene may indeed be able to be synthesized but the degree of defects and impurities in the grown films make them unsuitable for implementation. Figure 3.4 shows the Raman spectrum and in Figure 3.5 a transmission micrograph of few-layer graphene synthesized on SiO_2 using a plasma. The large D peak shows that the structure indeed contains a lot of defects. This work also demonstrated that the growth could be conducted at ambient temperatures with the only source of heating a plasma which elevated temperatures to ~300°C.

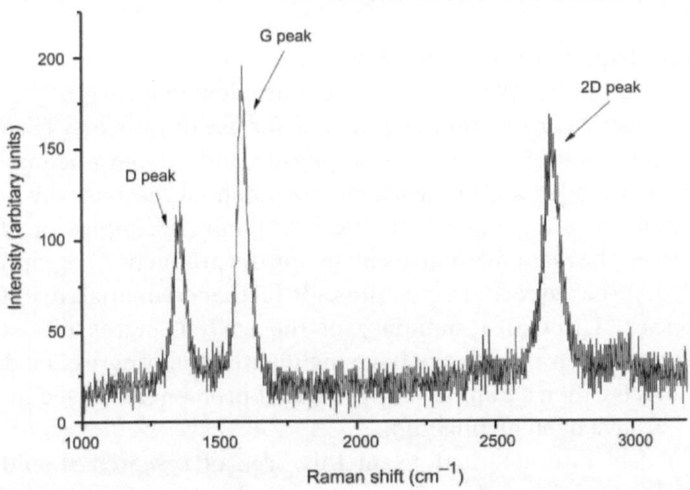

Figure 3.4 Raman spectrum of defective few-layer graphene grown without a catalyst and without heating, grown on SiO_2 with a methane plasma.

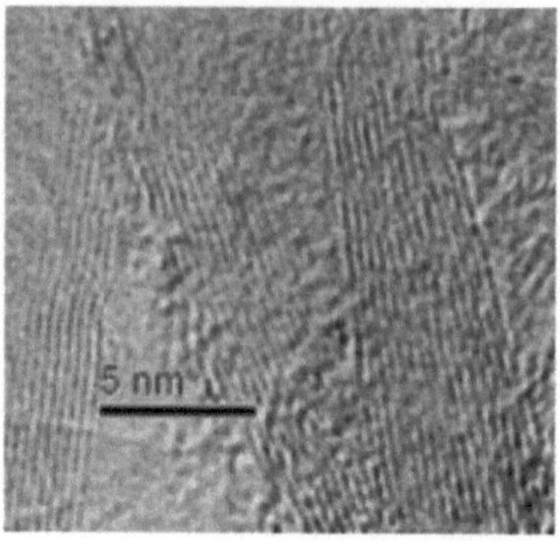

Figure 3.5 Transmission electron micrograph, average thickness appears to be around 7 layers with inter-planar distance of 0.34 nm corresponding well with few-layer graphene. Noticeable ridges and gaps consistent with the defective structure indicated by Raman spectrum.

Considering the infancy of investigations into graphene growth using PECVD the results obtained so far are quite interesting. More important is that the current investigations show a lot of promise for this method of synthesis for producing graphene in the future.

3.5 Concluding Remarks

In this chapter we have discussed the production of graphene in relation to PECVD. We have shown the problem that currently exists in the effective production of graphene for use by industry. Through an examination of simple plasma physics and related phenomena, we have provided a framework for the potential reasons why using a plasma may solve some of these problems. Examples of other successes plasmas have had in enhancing the production of graphene related sp^2-based carbon structures add further credentials to PECVD processes. Through a summary of the current status of PECVD production of graphene, we have highlighted the progress made in the field. Further, a number of the current problems involved in this process have been highlighted.

We feel future directions in this area of research should be related to examining the effect the plasma has on graphene growth. This will include but is not limited to examining which plasma systems are most effective; investigating the plasma parameters that have the greatest effect on growth; studying the interaction between surfaces and plasmas to enhance the growth of graphene; and developing a comprehensive picture for the growth mechanism of graphene. We hope that through this, an efficient approach can be found to producing graphene that is compatible with the industry's demands.

References

1. Schedin, F., Geim, A., Morozov, S., Will, E., Blake, P., Katsnelson, M., and Novoselov, K., (2007). Detection of individual gas molecules adsorbed on graphene, *Nature Materials*, **6**, p. 652–655.
2. Zhang, Y., Chen, Y., Zhou, K., Liu, C., Zeng, J., Zhang, H., and Peng, Y., (2009). Improving gas sensing properties of graphene by introducing dopants and defects: a first-principles study, *Nanotechnology*, **20**, p. 185504.
3. Lu, Y., Goldsmith, R., Kybert, N., and Johnson, A., (2010). DNA-decorated graphene chemical sensors, *Applied Physics Letters*, **97**, p. 083107.
4. Lundstrom, M., (2011). Graphene: The long and winding road, *Nature Materials*, **10**, p. 566–567.
5. Chitara, B., Panchakarla, L., Krupanidhi, S., and Rao, C., (2011). Infrared photodetectors based on reduced graphene oxide and graphene nanoribbons, *Advanced Materials*, p. DOI: 10.1002/adma.201101414.
6. Saha, S., Bakey, M., and Majumdar, D., (2010). Graphene quantum sheets: A new material for spintronic applications, *Advanced Materials*, **22**(48), p. 5531–5536.
7. Miller, J., Outlaw, R., and Holloway, B., (2010). Graphene double-layer capacitor with ac line-filtering performance, *Science*, **329**(5999), p. 1637–1639.
8. Mermin, N., (1968). Crystalline order in two dimensions, *Physical Review*, **176**, p. 250–254.
9. Novoselov, K., Geim, A., Morozov, S., Jiang, D., Zhang, Y., Dubonos, S., Grigorieva, I., and Firsov, A., (2004). Electric field effect in atomically thin carbon films, *Science*, **306**, p. 666–669.
10. Casiraghi, C., Hartschuh, A., Lidorikis, E., Qian, H., Harutyunyan, H., Gokus, T., Novoselov, K., and Ferrari, A., (2007). Rayleigh imaging of graphene and graphene layers, *Nano Letters*, **7**(9), p. 2711–2717.
11. Geim, A. and Novoselov, K., (2007). The rise of graphene, *Nature Materials*, **6**, p. 183–191.

12. Zhang, Y., Small, J., Pontius, W., and Kim, P., (2005). Fabrication and electric-field-dependent transport measurements of mesoscopic graphite devices, *Applied Physics Letters*, **86**, p. 073104.
13. Sidorov, A., Yazdanpanah, M., Jalilian, R., Ouseph, P., Cohn, R., and Sumanasekara, G., (2007). Electrostatic deposition of graphene, *Nanotechnology*, **18**, p. 4.
14. Soldano, C., Mahmood, A., and Dujardin, E., (2010). Production, properties and potential of graphene, *Carbon*, **48**, p. 2127–2150.
15. Hummers, W. and Offeman, R., (1958). Preparation of graphitic oxide, *Journal of the American Chemical Society*, **80**, p. 1339.
16. Mohanty, N., Nagaraja, A., Armesto, J., and Berry, V., (2009). High-throughput, ultrafast synthesis of solution-dispersed graphene via a facile hydride chemistry, *Small*, **6**, p. 226–231.
17. Li, Z., Yao, Y., Lin, Z., Moon, K., Lin, W., and Wong, C., (2010). Ultrafast, dry microwave synthesis of graphene sheets, *Journal of Materials Chemistry*, **20**, p. 4781–4783.
18. Stankovich, S., Dikin, D., Piner, R., Kohlhaas, K., Kleinhammes, A., Jia, Y., Wu, Y., Nguyen, S., and Ruoff, R., (2007). Synthesis of graphene-based nanosheets via chemical reduction of exfoliated graphite oxide, *Carbon*, **45**, p. 1558–1565.
19. Hiebel, F., Mallet, P., Magaud, L., and Veuillen, J., (2009). Atomic and electronic structure of monolayer graphene on 6H-SiC(0001)(3x3): A scanning tunneling microscope study, *Physical Review B*, **80**, p. 235429.
20. Virojanadara, C., Zakharov, A., Yakimova, R., and L., J., (2010). Buffer layer free large area bi-layer graphene on SiC(0001), *Surface Science*, **604**, p. L4–L7.
21. Robinson, J., Weng, X., Trumbull, K., Cavalero, R., Wetherington, M., Frantz, E., LaBella, M., Hughes, Z., Fanton, M., and Snyder, D., (2010). Nucleation of epitaxial graphene on SiC(0001), *ACS Nano*, **4**, p. 153–158.
22. Terai, M., Hasegawa, N., Okusawa, M., Otani, S., and Oshima, C., (1998). Electronic states of monolayer micrographite on TiC(111)-faceted and TiC(410) surfaces, *Applied Surface Science*, **130–132**, p. 876–882.
23. Itchkawitz, B., Lyman, P., Ownby, G., and D., Z., (1994). Monolayer graphite on TaC(111): electronic band structure, *Surface Science*, **318**, p. 395–402.
24. Sutter, P., Sadowski, J., and Sutter, E., (2009). Graphene on Pt(111): Growth and substrate interaction, *Physical Review B*, **80**, p. 245411.

25. Sutter, P., Fledge, J., and Sutter, E., (2008). Epitaxial growth of graphene on ruthenium, *Nature Materials*, **7**, p. 406–411.

26. Gruenis, A., Kummer, K., and Vyalikh, D., (2009). Dynamics of graphene growth on a metal surface: A time-dependent photoemission study, *New Journal of Physics*, **11**, p. 073050.

27. Coraux, J., N'Diaye, A., Engler, M., Busse, C., Wall, D., Buckanie, N., Meyer zu Heringdorf, F., van Gastel, R., Poelsema, B., and Michely, T., (2009). Growth of graphene on Ir(111), *New Journal of Physics*, **11**, p. 023006.

28. Preobrajenski, A., Ng, M., Vinogradov, A., and Martensson, N., (2008). Controlling graphene corrugation on lattice mismatched substrates, *Physical Review B*, **78**, p. 073401.

29. Eom, D., Prezzi, D., Rim, K., Zhou, H., Lefenfeld, M., Xiao, S., Nuckolls, C., Hybersen, M., Heinz, T., and Flynn, G., (2009). Structure and electronic properties of graphene nanoislands on Co(0001), *Nano Letters*, **9**, p. 2844–2848.

30. Reina, A., Thiele, S., Jia, X., Bhaviripudi, S., Desselhaus, M., Schaefer, J., and Kong, J., (2009). Growth of large-area single- and bi-layer graphene by controlled carbon precipitation on polycrystalline Ni surfaces, *Nano Research*, **2**, p. 509–516.

31. Kang, B., Mun, J., Hwang, C., and Cho, B., (2009). Monolayer graphene growth on sputtered thin film platinum, *Journal of Applied Physics*, **106**, p. 104309.

32. Wang, X., You, H., Liu, F., Li, M., Wan, L., Li, S., Li, Q., Xu, Y., Tian, R., Yu, Z., Xiang, D., and Cheng, J., (2009). Large-scale synthesis of few-layered graphene using CVD, *Chemical Vapor Deposition*, **15**, p. 53–56.

33. Li, X., Cai, W., An, J., Kim, S., Nah, Yang, D., Piner, R., Velamakanni, A., Jung, I., Tutuc, E., Banerjee, S., Colombo, L., and Ruoff, R., (2009). Large-area synthesis of high-quality and uniform graphene films on copper foils, *Science*, **324**, p. 1312–1314.

34. Mariotti, D. and Sankaran, R., (2010). Microplasmas for nanomaterials synthesis, *Journal of Physics D: Applied Physics*, **43**, p. 323001.

35. Cvelbar, U. and Mozetic, M., (2007). Behaviour of oxygen atoms near the surface of nanostructured Nb_2O_5, *Journal of Physics D: Applied Physics*, **40**, p. 2300–2303.

36. Cvelbar, U., Chen, Z., Sunkara, M., and M., M., (2008). Spontaneous growth of superstructure α-Fe_2O_3 nanowire and nanobelt arrays in reactive oxygen plasma, *Small*, **4**, p. 1610–1614.

37. Tsakadze, Z., Levchenko, I., Ostrikov, K., and Xu, S., (2007). Plasma-assisted self-organized growth of uniform carbon nanocone arrays, *Carbon*, **45**, p. 2022–2030.

38. Teo, K., Hash, D., Lacerda, R., Rupesinghe, N., Bell, M., Dalal, S., Bose, D., Govindan, T., Cruden, B., Chhowalla, M., Amaratunga, G., Meyyappan, M., and Milne, W., (2004). The significance of plasma heating in carbon nanotube and nanofiber growth, *Nano Letters*, **4**, p. 921–926.

39. Perrin, J., Shiratani, M., Kae-Nune, P., Videlot, H., Jolly, J., and Guillon, J., (1998). Surface reaction probabilities and kinetics of H, SiH_3, Si_2H_5, CH_3, and C_2H_5 during deposition of *a*-Si:H and *a*-C:H from H_2, SiH_4, and CH_4 discharges, *Journal of Vacuum Science and Technology A*, **16**, p. 278–289.

40. Das Arulsamy, A. and Ostrikov, K., (2009). Diffusivity of adatoms on plasma-exposed surfaces determined from the ionization energy approximation and ionic polarizability, *Physics Letters A*, **373**, p. 2267–2272.

41. Mariotti, D. and Ostrikov, K., (2009). Tailoring microplasma nanofabrication: from nanostructures to nanoarchitectures, *Journal of Physics D: Applied Physics*, **42**, p. 092002.

42. Sankaran, R.M., (2011). Towards chirality-controlled SWCNTs: can a plasma help, *Journal of Physics D: Applied Physics*, **44**, p. 174005.

43. Nozaki, T., Yoshida, S., Karatsu, T., and Okazaki, K., (2011). Atmospheric-pressure plasma synthesis of carbon nanotubes, *Journal of Physics D: Applied Physics*, **44**, p. 174007.

44. Melecheko, A., Pearce, R., Hensley, D., Simpson, M., and McKnight, T., (2011). Challenges in process integration of catalytic DC plasma synthesis of vertically aligned carbon nanofibres, *Journal of Physics D: Applied Physics*, **44**, p. 174008.

45. Keidar, M., Shashurin, A., Li, J., Volotskova, O., Kundrapu, M., and Zhuang, T.S., (2011). Arc plasma synthesis of carbon nanostructures: where is the frontier?, *Journal of Physics D: Applied Physics*, **44**, p. 174006.

46. Hatakeyama, R., Kaneko, T., Kato, T., and Li, Y.F., (2011). Plasma-synthesized single-walled carbon nanotubes and their applactions, *Journal of Physics D: Applied Physics*, **44**, p. 174004.

47. Tsugawa, K., Ishihara, M., Kim, J., Koga, Y., and Hasegawa, M., (2010). Nanocrystalline diamond film growth on plastic substrates at temperatures beloew 100°C from low-temperature plasma, *Physical Review B*, **82**, p. 125460.

48. Bower, C., Zhou, O., Zhu, W., Werder, D., and Jin, S., (2000). Nucleation and growth of carbon nanotubes by microwave plasma chemical vapor deposition, *Applied Physics Letters*, **77**, p. 2767–2769.

49. Tam, E. and Ostrikov, K., (2008). Plasma-controlled adatom delivery and (re)distribution: Enabling uninterrupted, low-temperature growth

of ultralong vertically aligned single walled carbon nanotubes, *Applied Physics Letters*, **93**, p. 261504.

50. Hofmann, S., Ducati, C., Robertson, J., and Kleinsorge, B., (2003). Low-temperature growth of carbon nanotubes by plasma-enhanced chemical vapor deposition, *Applied Physics Letters*, **83**, p. 135–137.

51. Denysenko, I., Ostrikov, K., Yu, M., and Azarenkov, N., (2007). Effects of ions and atomic hydrogen in plasma-assisted growth of single-walled carbon nanotubes, *Journal of Applied Physics*, **102**, p. 074308.

52. Baker, R. and Barber, M., (1978). *Chemistry and Physics of Carbon, Volume 14* (Marcel Dekker, New York), p. 83.

53. Meyyappan, M., Delzeit, L., Cassell, A., and Hash, D., (2003). Carbon nanotube growth by PECVD: a review, *Plasma Sources Science and Technology*, **12**(2), p. 205–216.

54. Hiramatsu, M., Shiji, K., Amano, H., and Hori, M., (2004). Fabrication of vertically aligned carbon nanowalls using capacitively coupled plasma-enhanced chemical vapor deposition assisted by hydrogen radical injection, *Applied Physics Letters*, **84**, p. 4708–4710.

55. Wu, Y., Qiao, P., Chong, T., and Shen, Z., (2002). Carbon nanowalls grown by microwave plasma enhanced chemical vapor deposition, *Advanced Materials*, **14**, p. 64–67.

56. Levchenko, I., Ostrikov, K., Rider, A., Tam, E., and S., V., (2007). Growth kinetics of carbon wall-like structures in low temperature plasmas, *Physics of Plasmas*, **14**, p. 063502.

57. Chuang, A., Boskovic, B., and Robertson, J., (2006). Freestanding carbon nanowalls by microwave plasma-enhanced chemical vapour deposition, *Diamond & Related Materials*, **15**, p. 1103–1106.

58. Zhu, M., Wang, J., Outlaw, R., Hou, K., Manos, D., and Holloway, B., (2007). Synthesis of carbon nanosheets and carbon nanotubes by radio frequency plasma enhanced chemical vapor deposition, *Diamond & Related Materials*, **16**(196), p. 196–201.

59. Nandamuri, G., Roumimov, S., and Solkani, R., (2010). Remote plasma assisted growth of graphene films, *Applied Physics Letters*, **96**, p. 154101.

60. Qi, J.L., Zheng, W.T., Zheng, X.H., Wang, X., and Tian, H.W., (2011). Relatively low temperature synthesis of graphene by radio frequency plasma enhanced chemical vapor deposition, *Applied Surface Science*, **257**, p. 6531–6534.

61. Wang, S.M., Pei, Y.H., Wang, X., Meng, Q.N., Tian, H.W., Zheng, X.L., Zheng, W.T., and Liu, Y.C., (2011). Synthesis of graphene on a polycrystalline Co

film by radio-frequency plasma-enhanced chemical vapor deposition, *Journal of Physics D: Applied Physics*, **43**, p. 455402.

62. Kim, S., Ishihara, M., Koga, Y., Tsugawa, K., Hasegawa, M., and Iijima, S., (2011). Low-temperature synthesis of large-area graphene-based transparent conductive films using surface wave plasma chemical vapor deposition, *Applied Physics Letters*, **9**, p. 091502.
63. Lee, B., Lee, T., Park, S., Yu, H., Lee, J., Lim, S., and Jeong, G., (2011). Low-temperature synthesis of thin graphite sheets using plasma-assisted thermal vapor deposition system, *Materials Letters*, **65**, p. 1127–1130.
64. Zhang, L., Shi, Z., Wang, Y., Yang, R., Shi, D., and Zhang, G., (2011). Catalyst-free growth of nanographite films on various substrates, *Nano Research*, **4**(3), p. 315–321.

Chapter 4

Morphology Control at the Nanoscale of Polymer Composites for Organic Solar Cells

Bernard Ratier,[a,*] Johann Bouclé,[a] Ghislain Boloma,[a] Fabrice Rossignol,[b] and Valérie Coudert[b]

[a] *XLIM Research Institute, CNRS UMR 6172, Limoges University, 123 Av Albert Thomas, 87060 Limoges, France*

[b] *SPCTS, CNRS UMR 6638, Limoges University, 12 rue Atlantis, 87068 Limoges, France*

*bernard.ratier@unilim.fr

Organic solar cells can offer flexibility, light weight, semitransparency and large-area applications which may be regarded as an attractive alternative to their inorganic counterparts. Such devices are based on an active layer made of a polymer-nanoparticle composite, where the polymer is an electron donor material and the nanoparticles form an electron acceptor material. Device operation is based on the photoinduced charge transfer between the absorbing donor polymer and the acceptor molecules. Taking into account the short exciton diffusion length in organic materials (10 nm range), the absorbing layer must be formed by a blend of the donor and the acceptor

Advances in Nanodevices and Nanofabrication: Selected Publications from Symposium of Nanodevices and Nanofabrication in ICMAT2011
Edited by Qing Zhang, W. I. Milne, and Jianping Zou

ISBN 978-981-4364-54-6 (Hardcover), 978-981-4364-55-3 (eBook)
www.panstanford.com

materials with domains in the range of this diffusion length. The fine control of polymer-nanoparticle composites at the nanoscale is a key parameter for improving organic solar cell performance. However, there is still a lack of a clear relationship between fabrication processes and the resulting morphology. In order to observe this morphology and to give further insight into cell operation, near-field microscopy experiments under illumination have been performed using Kelvin probe force microscopy (KPFM) under ultrahigh vacuum and current sensing (CAFM). High resolution images have been obtained which allow a realistic observation of the device operating at the nanometer scale.

4.1 Introduction

The sun provides a unique opportunity to satisfy the worldwide demand for low cost and clean energy. Conversion of sun light to electricity is achieved by the photovoltaic process. The majority of solar cells currently available on the market are rather expensive and relatively rigid when compared to polymer solar cells (PSCs). Because organic solar cells are based on thin-film technology with low energy payback, flexibility, light weight, semitransparency and large-area applications are available and they present an attractive alternative to their inorganic counterparts [1]. Such devices are based on an active layer made of a polymer-nanoparticle composite, where the polymer is an electron-rich material (electron donor) and the nanoparticles make up an electron-deficient material (electron acceptor). Device operation is based on the photoinduced charge transfer between the absorbing donor polymer and the acceptor molecule [2–4]. Charge carriers are photogenerated and subsequently separated at the donor-acceptor (DA) interface by exciton dissociation. Photons are absorbed by the polymer and the energy stimulates a coulombically bound electron-hole pair, known as the exciton, from the highest occupied molecular orbital (HOMO) to the lowest unoccupied molecular orbital (LUMO) of the polymer. This is followed by exciton dissociation due to electron transfer from the LUMO of the donor to the LUMO of the acceptor. Taking into account the short exciton diffusion length in organic materials (10 nm range), the absorbing layer must be formed by a blend of the donor and the acceptor materials with domains in the range of this diffusion length. Moreover, the blend has to form a crystalline percolating interpenetrating network (IPN) to insure charge transport to the electrodes. The fine control of polymer-nanoparticle composites at the nanoscale is a key parameter for improving organic solar cell performance. Such a challenging morphology can be obtained by optimizing all steps involved in cell fabrication, from solution formulation to cell annealing. The aim of this paper is to focus on the nanoscale morphology control of the polymer-fullerene active layer.

In PSCs, the IPN is obtained by blending a conjugated polymer (as the donor material) and a buckminsterfullerene derivative (as the acceptor), sandwiched between two electrodes. One of the electrodes is a transparent conductive oxide, most commonly Indium Tin Oxide (ITO), insuring efficient illumination and hole collection, while the other is generally made of a low work function metal like aluminum for electron collection. The most common buckminsterfullerene derivative is [6,6]-phenyl-C61-butyric acid methyl ester ([C60] PCBM or PCBM) which is miscible with conjugated polymers, a highly efficient electron acceptor with a relatively high electron mobility [5]. Many electron donor conjugated polymers are used in PSC studies with the most common being poly(phenylenevinylenes), polyfluorenes, polycarbazoles and polythiophenes [6]. Nowadays, the most commonly available polymer is the regioregular poly-3(hexylthiopene) r-(P3HT) [7], which will be considered a model system in the present studies. This polymer blended with PCBM led to power conversion efficiencies up to the 5% [8–11]. In order to complete the device, a hole transport layer can be introduced between the anode and the active layer, most commonly the poly(3,4-ethylenedioxythiophene) poly(styrenesulfonate) (PEDOT-PSS). Figure 4.1 provides the device architecture and chemical structure of the materials used.

In the first part, a correlation between the process conditions during the active layer fabrication and the resulting P3HT-PCBM based solar cell efficiency will be established. In order to verify that the proper morphology is obtained, and to give further insight into cell operation, near-field microscopy experiments in the dark and under illumination have been performed using Kelvin probe force microscopy (KPFM) under ultrahigh vacuum and atomic force microscopy in current sensing mode (CAFM). High resolution images have been obtained giving insight into device operation under illumination.

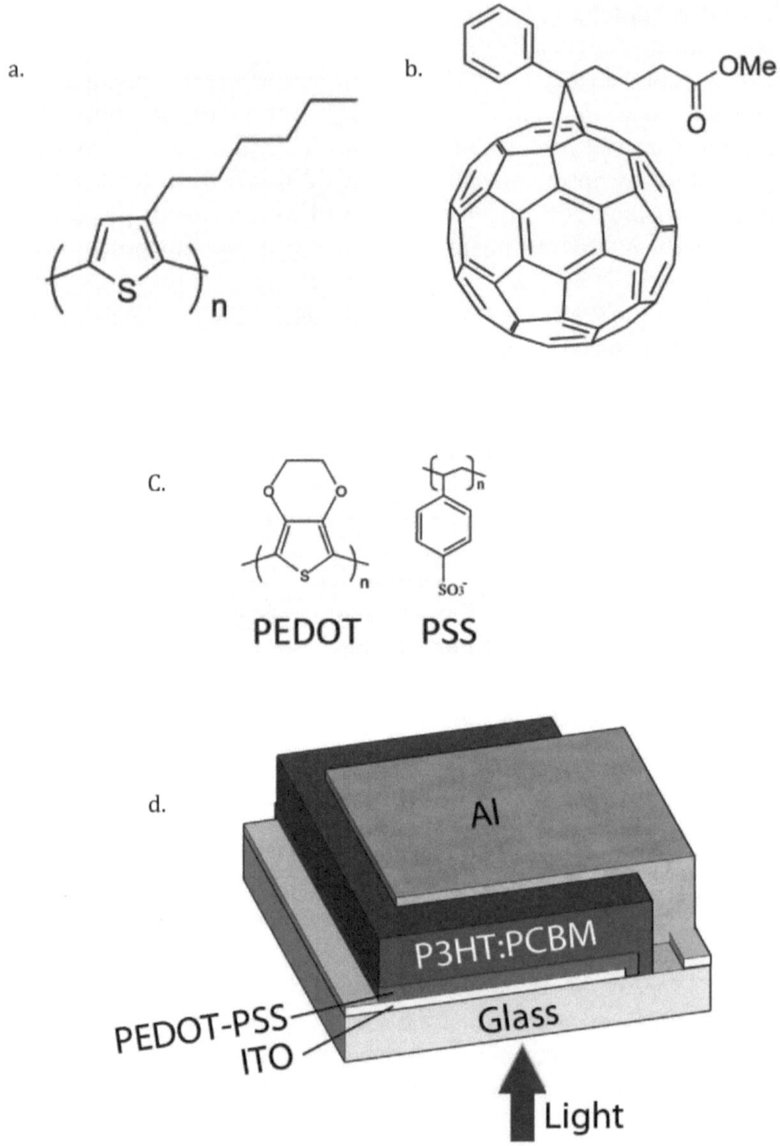

Figure 4.1 Chemical structures of the most common materials used in PSCs: (a) donor conjugated polymers: poly(3-hexylthiophene) (P3HT). (b) acceptor fullerene derivative: [6,6]-phenyl-C61-butyric acid methyl ester (PCBM). (c) hole transport layer: poly(3,4-ethylenedioxythiophene) poly(styrenesulfonate) (PEDOT:PSS). (d) polymer solar cell architecture.

4.2 P3HT-PCBM-Based Solar Cells

4.2.1 Main Photovoltaic Parameters

Figure 4.2 shows the *I–V* curve of a solar cell under illumination. The power conversion efficiency (PCE) can be expressed by equation (1) using the measured values of the short circuit current I_{SC} and open circuit voltage V_{OC} when the cell is under solar illumination, and the maximum power point (I_m, V_m) of the *I–V* curve [4]. The standard illumination conditions correspond to the so-called AM 1.5 G solar spectrum at 100 mW cm^{-2} incident irradiance, matching the sun spectral intensity distribution on the earth's surface when it is tilted at an angle of 48.2° with respect to the normal direction.

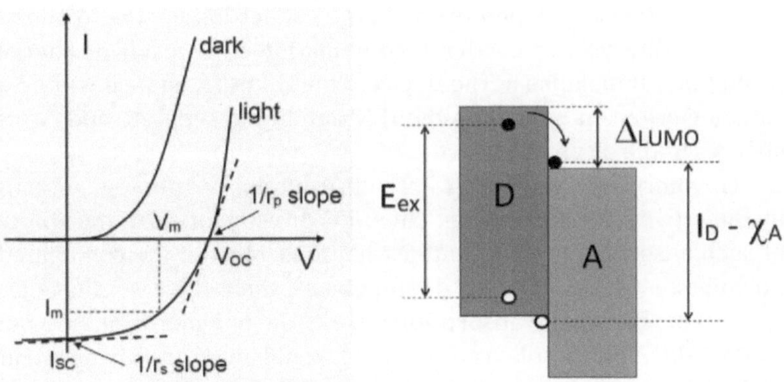

Figure 4.2 (a) Current versus voltage curve in the dark and under illumination for a solar cell showing the main photovoltaic parameters: open circuit voltage V_{OC}, short circuit current I_{SC}. Power conversion efficiency and fill factor FF are deduced from equation (1). (b) Charge transfer mechanism between donor and acceptor materials. Blue arrows indicate the right way to improve the V_{OC} or the polymer absorption, and red arrow illustrates the way to improve absorption which is detrimental to the V_{OC}.

$$PCE = \frac{\text{Maximal power produced per surface unit}}{\text{Total incident light power per surface unit}}$$

$$= \frac{P_m}{\Phi_{in}} = \frac{V_m I_m}{\Phi_{in}} = \frac{FF.I_{SC}V_{OC}}{\Phi_{in}} \quad (4.1)$$

with FF = $(I_m V_m) / (V_{OC} I_{SC})$ being the fill factor of the cell and Φ_{in} the incident light energetic flux.

Another useful data will be the external quantum efficiency (EQE) which can be expressed at a given wavelength by the ratio between the number of collected charges and the number of incident photons per second.

From the PCE expression, one can maximize the solar cell performance by maximizing the values of V_{OC}, I_{SC}, and FF.

The FF parameter depends on the cell technology and on the active layer morphology. As it can be seen in Figure 4.2, the I–V curves under illumination will show an ideal characteristic if the series resistance of the device r_s is close to 0 and shunt resistance r_p approaches an infinite value [12]. The series resistance is attributed to both material resistivity and contact resistance of the cell. The shunt resistance r_p depends on charge carrier losses due to short-circuit pathways and carrier recombination as a result of charge or mobility imbalance in the device. This shunt resistance will be a valuable criterion indicating the IPN's ability to separate and carry charges to the electrodes.

The short circuit current, I_{SC} results from charges photogenerated by the active layer absorption. Thus, it is directly linked to the ability of such absorbing layer to harvest photons. Most of the conjugated polymers used in solar cells do not absorb the red part of the solar spectrum. Shifting the absorption edge of the polymer from 550 nm (MDMO-PPV: PCBM solar cells) to 1 μm would imply an enhancement of the harvested photons from 25% to 90% [13].

The open circuit voltage, V_{OC} was found to mainly depend on the respective energy levels of the donor and acceptor materials. Exciton separation into free charges at the DA interface of the PSC can be understood as a charge transfer of the electron from the LUMO of the donor to the LUMO of the acceptor. The gain in energy (Δ_{LUMO}) gives rise to the driving force necessary for charge separation (Figure 4.2b). The separated charges will reach the LUMO of the acceptor (electron affinity χ_A) and the HOMO of the donor (ionization potential I_D) for electrons and holes respectively. Due to polymer lattice distortion, the exciton binding energy E_{ex} is somewhat lower than the donor band gap, so that the separation condition will be $E_{ex} > I_D - \chi_A$. In the case of the P3HT-PCBM solar cells (Δ_{LUMO} = 0.7 eV), the exciton separation efficiency is assumed to approach 1. Considering electrodes with ohmic contact, the V_{OC} can be expressed as [14]

$$V_{OC} = (1/e)\,(|\,E_{\mathrm{HOMO}}^{\mathrm{donor}}\,| - |\,E_{\mathrm{LUMO}}^{\mathrm{acceptor}}\,|) - \Delta V, \qquad (4.2)$$

where 0.2 V < ΔV < 0.5 eV depending on the nature of the polaron states in the material [15].

4.2.2 Correlation Between Device Processing and Efficiency

As mentioned in the introduction, the IPN morphology of the active layer plays a key role in building efficient devices. The ideal IPN should meet the following criteria: (i) a maximum interfacial area to favor exciton dissociation, (ii) a domain size of the order of magnitude of the exciton diffusion length, (iii) crystallization of the two compounds of the IPN should be effective so as to favor high mobility charge transport, charge balance between holes and electrons and to limit charge recombination, and (iv) the IPN must be bi-continuous, which means the different domains of the two materials have to be linked to create percolation pathways to the electrodes. The absence of some or all of the above features can seriously affect the cell's fill factor. Processing conditions influencing the morphology are mainly the solvent used in the blend formulation and thermal annealing. An important milestone for achieving large-area PSC production will be performance reproducibility, which in turn requires reproducible fabrication processes and material supply: in order to fabricate a Giga Watt power plant made of organic solar cells, and taking into account cell geometry, polymer and fullerene derivative need to be produced by the ton. As it is very tricky to obtain a polymer reproducibility at the ton scale, an important question must be resolved: Is it possible to predict an optimized process taking into account physical and chemical material discrepancies?

The effect of the solvent has been clearly established in the case of MDMO-PPV:PCBM PSCs where toluene was replaced by dichlorobenzene (DCB), doubling the device efficiency [16]. As observed by SEM and AFM experiments, the improved efficiency is attributed to the different morphologies obtained in both cases, where the use of toluene results in a large domain size and non bi-continuous IPN, while DCB allows a ten-fold decrease of the domain size [17]. Another fruitful strategy consists of adding a high boiling point selective solvent (i.e. in which only PCBM is soluble) to the solution [18]. In this case, P3HT crystallization occurs more rapidly, and a PCBM liquid phase can more easily interpenetrate the P3HT. The resulting morphology is improved where more interconnected and crystallized fullerene domains in the IPN are observed, improving electron transport in the cell.

In all cases, the resulting IPN morphology originates from the non-miscible nature of the polymer and the fullerene. This is where the deposition process (from solution formulation to annealing of the active layer) has a crucial influence: spin coating at various speeds and durations will yield different morphologies. The solvent annealing process, which leads to phase separation after deposition of the solution under a saturated solvent atmosphere, affects the spinodal decomposition in the liquid phase and can yield efficient devices [19].

However, thermal annealing remains the last step of the phase separation process where the molecular weight, regioregularity and polydispersity have an important role to play, which means that the appropriate annealing parameters are polymer-dependent. It has been shown that in the case of P3HT and P3BT cells [20], a high polydispersity and high molecular weight polymer generally results in the best performance but with no clear correlation between material properties and annealing temperature. Not surprisingly, different "optimized" annealing temperatures and times can be found in the literature, varying in most cases from 100°C to 150°C [8–11]. An example is given in Figure 4.3 in the case of a high molecular weight and highly regioregular polythiophene, with a common 1:0.8 P3HT-PCBM ratio [21]. Annealing was applied after device fabrication during 10 minutes under inert atmosphere. The initial solar cell without annealing have a PCE of 2.2% (Figure 4.3a), which can be optimized to 4% with annealing at 100°C. The corresponding external quantum efficiencies provided in As shown in Figure 4.3b, the main annealing effect at such an optimized low temperature is to increase the PCBM contribution to the overall photocurrent generation while the polymer structure remains unchanged, since the annealing temperature is below the P3HT glass transition temperature. The effect of annealing has to be understood here as to favor diffusion of PCBM into the polymer in order to create electron percolating pathways.

In the case of higher annealing temperatures, around the glass transition temperature of the polymer, the most commonly agreed upon mechanisms occurring during thermal annealing are

(i) enhancement of the polymer crystallization into larger grains, (ii) PCBM diffusion between P3HT crystallites forming needles, and (iii) formation of the IPN.

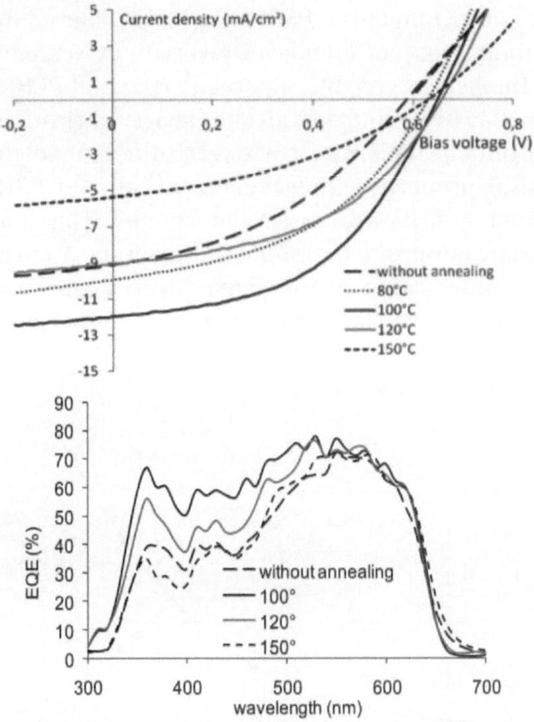

Figure 4.3 (a) Current density versus bias voltage under AM 1.5 illumination (100 mW/cm^2) of P3HT-PCBM PSCs for different annealing temperatures. (Active layer blend obtained with 1:0.8 P3HT–PCBM ratio in dichlorobenzene). (b) Corresponding external quantum efficiencies.

Consequently, it may be possible to control the phase separation kinetics by changing the material's concentration [21,22]. Such an effect is clearly demonstrated in Figure 4.4, which represents the *I–V* characteristics under AM1.5 illumination of solar cells made with the same P3HT-PCBM weight ratio (1:0.8), the same annealing treatment, but with increasing material concentrations in the liquid phase prior to film deposition. The active layer thickness of the devices was kept constant (100 nm) for comparison. Power conversion efficiencies vary from 0.8% to 4.3% by tripling the material concentration. We have proposed a decomposition kinetic model in which very high material concentrations in the liquid phase limit the free space between moieties and slow down molecular rearrangement, giving rise to a smaller segregated domain size [21]. Currently, a large number of PSCs have been achieved by the Plastic Optoelectronic group of Limoges University by screening different polymers (molecular weight, regioregularity) and PCBM molecules. It appears that by optimizing all the device fabrication steps, it is possible to obtain efficient cells with very different polymer physical and chemical properties. However, each batch of P3HT needs to be optimized, and its properties determined. This effort is being undertaken by our group to predict these optimization conditions.

A direct observation of the active layer nanostructure is thus mandatory in order to clarify the process–morphology relationship.

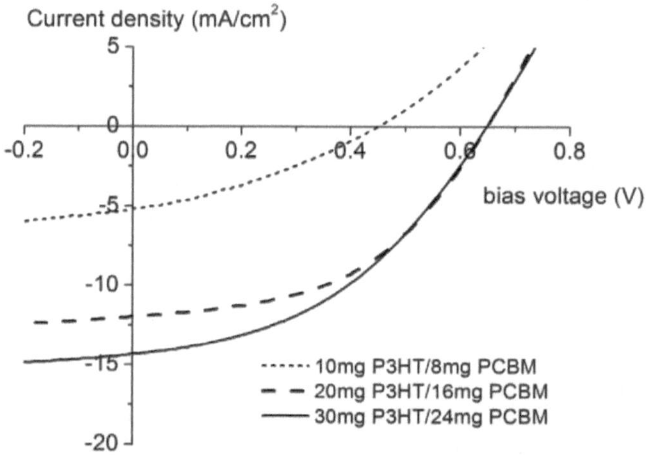

Figure 4.4 Current density versus bias voltage under AM 1.5 illumination (100 mW/cm^2) of P3HT-PCBM PSCs for different material concentrations (active layer blend obtained with 1:0.8 P3HT-PCBM ratio in dichlorobenzene).

4.2.3 Nanoscale Characterization of the Blend Structure and Carrier Photogeneration Using Near-Field Microscopy

As mentioned above, the nanoscale morphology of the active layer plays a key role in the resulting device efficiency. A challenging objective is to observe directly this morphology, since the sensitivity of the experiment has to be around the nanometer range. Near-field microscopy experiments can achieve this objective. A first approach consists of the direct observation of the structure, using for example the tapping mode of an atomic force microscopy (AFM) to investigate the topography of the active layer and to provide surface roughness. An example is given in Figure 4.5 for two of the devices that have been characterized in Figure 4.4. It is clearly shown here that doubling the material concentration in the solution phase can produce very different structures before annealing, with a grain size that is significantly different (up to one order of magnitude) depending on the material concentration. However, it is not possible to deduce from these experiments that the observed structure corresponds to a donor or acceptor domain size.

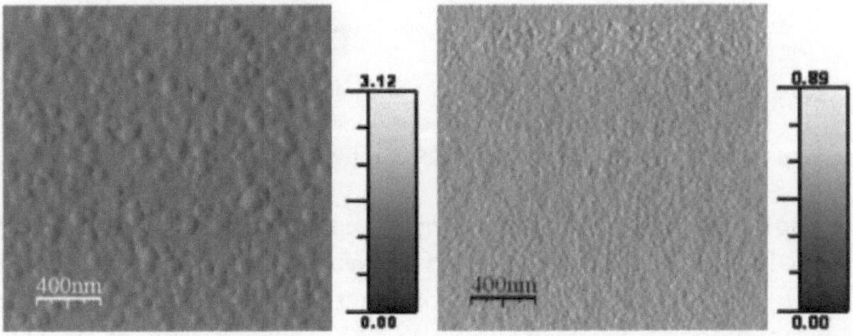

Figure 4.5 AFM topography of two P3HT-PCBM active layers processed from solutions in dichlorobenzene with different material concentrations: in the left image the solution is 10 mg P3HT/mL dichlorobenzene:8 mg PCBM/mL dichlorobenzene, while in the right image this solution is 20 mg P3HT/mL dichlorobenzene:16 mg PCBM/mL dichlorobenzene.

A better understanding of device operation is expected if the dependence of photocurrent generation upon active layer morphology and structure can be analyzed in detail. In this context, several studies reported the use of conductive and photo-conductive atomic force microscopy (CAFM and pCAFM respectively) to locally map the photocurrent of organic bulk heterojunctions in the dark or under illumination [23–26]. While electrostatic force microscopy (EFM) and scanning Kelvin probe microscopy (SKPFM) techniques can be used to perform local potentiometry at the surface with a relatively high resolution [26,27], CAFM gives information mainly on the vertical charge transport properties of a donor-acceptor blend. In CAFM, a voltage bias is applied between a conductive tip at the surface of the sample and a conductive substrate. In the case of an organic solar cell, the tip is used as the collecting electrode for either holes or electrons, depending on the work function of the metallic coating, with the transparent conductive anode closing the electrical circuit (Figure 4.6a).

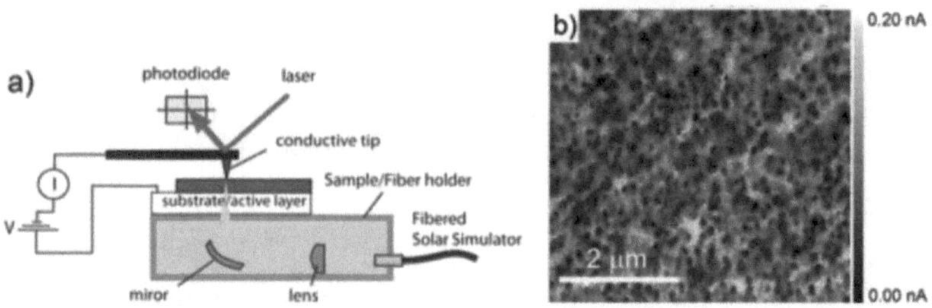

Figure 4.6 Schematic experimental photo-conductive AFM set-up (a) and typical current map (b) obtained in the dark on a P3HT:PCBM organic active layer under applied bias (–0.5 V).

Both the current and topography can then be simultaneously recorded as the tip scans the surface, with a high lateral resolution that can reach the exciton diffusion length (~20 nm). A typical dark current map of a cell based on P3HT:PCBM (weight ratio of 1:1.2) is shown in Figure 4.6b. The use of an electrical system with a broad dynamic range allows a fine discrimination between conducting and non-conducting regions. The technique has been applied extensively on polymer/fullerene heterojunctions, enabling the determination of the composition of phase separated domains [23]. Additionally, it is possible to record current-voltage characteristics at the nanoscale, and to assess or possibly quantify charge carrier mobility associated with specific local features of active blends [28,29]. Initially implemented in the dark, the technique was more recently adapted using various light sources such as laser beams in order to monitor the electrical changes that occur at the nanoscale upon irradiation [24,25]. In particular, the evolution of spatially resolved photocurrent as a function of active layer treatment was reported on polythiophenes/fullerene blends, showing that heterogeneous charge generation and transport occur even in very efficient cells

[30,31]. The versatility of the technique is also demonstrated in organic solar cells based on small-molecular weight based organic solar cells [32] or using polymer nanowires [33]. The technique is of course possible if there is no interaction between the laser beam probing the cantilever displacement and the device illumination beam. This is usually always the case for organic solar cells since the wavelength of the AFM laser is usually located well beyond the absorption spectrum of the active layer of the cell under test. In all cases, it has been demonstrated that pCAFM is a powerful technique for simultaneously extracting topography images, dark and photo-current maps, as well as hole and electron currents, giving a broad picture of the electrical features at the nanoscale of organic active layers. In particular, the technique is found to be suitable for addressing morphological issues relating the active layer microstructure to the overall device performance.

However non-contact AFM techniques remain to date the more accurate tools to provide high lateral resolution in the nanometer range when operated under ultra-high vacuum. The ability to perform on the same sample, AFM by frequency modulation, the topography, the energy dissipation (damping) and the surface potential (by Kelvin probe force microscopy KPFM) cartography has allowed the INAC/SPrAM team from the CEA (Grenoble, France) to achieve the best resolved surface images of a P3HT-PCBM active layer blend to date [34]. The images obtained on high efficiency solar cells provided by our group [21] revealed domain sizes around 15 nm in the blend, which is in the exciton diffusion length range. The damping mode is a useful tool, which can reveal the nature of the surface since the energy dissipation resulting from the tip-surface interaction depends on the material's rigidity. By comparison with the topography, it is then possible to identify the nature of the domains. The images have been performed in dark and under illumination conditions at a wavelength which corresponds to P3HT absorption. By comparing images of the surface potential in the dark and under-illumination measured by KPFM, it was possible to reveal the surface photovoltage with a 2 nm resolution. These images clearly confirmed the existence of a very narrow space charge area between domains which can induce a very high internal field in an operating polymer solar cell, giving rise to probably a filamentary structure of the current driven in the cell.

4.3 Conclusion

Efficiency of polymer solar cells is greatly dependent on the active layer morphology at the nanoscale. This morphology can be controlled by conveniently tuning all the steps of the device fabrication process. However, this optimization greatly depends on the physical and chemical properties of the materials, and it remains challenging to predict the evolution of this optimization based on the material properties. However only annealing and solvent effect have been discussed here in order to obtain the required active layer morphology to obtain efficient devices. Including for example high anisotropic nanoparticles such as carbon nanotubes seems to be a promising method to help the active layer structuration and the charge carrier percolation to the electrodes [35].

Direct observation of the nanostructure can be obtained at the nanometer scale by near-field microscopy techniques, and constitutes a useful tool for understanding the relationship between morphology and cell performance. Moreover, additional knowledge of device operation can be obtained from direct observation of the photo-carrier generation under illumination. Nowadays, it is possible to directly observe the photocurrent and photovoltage in an organic solar cell at the nanometer scale and to have access to space charge dimension between donor and acceptor domains, or to observe the filamentary structure of the current.

Acknowledgments

The authors thank the CNRS Interdisciplinary Energy Program for funding (CELASOL Project). E. Spadafora, R. Demadrille and B. Grévin (INAC/SPrAM, UMR CNRS 5819, CEA-Grenoble, France) are gratefully acknowledged for their valuable KPFM measurements on our organic solar cells that are described in the last section.

References

1. Strategic Research Agenda (2009), *Organic & Large Area Electronics*, 2009, http://www.photonics21.org/uploads/OLAESRAFINALVS491509.pdf.
2. Gregg, B.A., (2003). In *Molecules as Components in Electronic Devices*, edited by M. Lieberman, American Chemical Society, Washington, DC, p. 243.

3. Brabec, C.J., Dyakonov, V., Parisi, J., Saricifci, N.S. (2003). *Organic Photovoltaics: Concepts and Realizations*, Springer-Verlag Berlin Heidelberg, New York.

4. Moliton, A. (2005). *Optoelectronics of Molecules and Polymers*, Springer Series in Optical Sciences, Springer-Verlag New York Inc.

5. Mihailetchi, V.D., Van Duren, J.K.J., Blom P.W.M., Hummelen, J.C., Janssen, R.A.J., Kroon, J.M., Rispens, M.T., Verhees, W.J.H., Wienk, M.M. (2003). Electron transport in a methanofullerene *Adv. Funct. Mater.* **13**, 43.

6. Cheng, Y.J., Yang, S.H., Hsu, C.S., (2009). Synthesis of conjugated polymers for organic solar cell applications, *Chem. Rev.*, **109**, 5868–5923.

7. Pandey, S.S., Takashima, W., Nagamatsu, S., Endo, T., Rikukawa, M., Kaneto, K. (2000). Regioregularity versus regiorandomness: effect on photocarrier transport in poly(3-hexylthiophene), *Japan. J. Appl. Phys.* **39,** L94–L97.

8. Reyes-Reyes, M., Kim. K., Dewald. J., Lopez-Sandoval, R., Avadhanula, A., Curran, S., Carroll, D.L. (2005). Meso-structure formation for enhanced organic photovoltaic cells, *Org. Lett.*, **7**: 26, 5749–5752.

9. Ma, W.L., Yang. C.Y., Gong, X., Lee, K., Heeger, A.J., (2005). Thermally stable, efficient polymer solar cells with nanoscale control of the interpenetrating network morphology, *Adv. Funct. Mater.*, **15**: 1617–1622.

10. Li, G., Shrotriya, V., Yao, Y., Yang, Y., (2005). Investigation of annealing effects and film thickness dependence of polymer solar cells based on poly(3-hexylthiophene, *J. App. Phys.*, **98**: 043704.

11. Chirvase D., Parisi J., Hummelen J.C., Dyakonov V. (2004). Influence of nanomorphology on the photovoltaic action of polymer–fullerene composites, *Nanotechnology*, **15**: 1317–1323.

12. Moliton, A., Nunzi, J.M. (2006). How to model the behaviour of organic photovoltaic cells, *Polym. Int.*, **55**, 583–600.

13. Sariciftci, N.S., (September 2004). Plastic photovoltaic devices, *Materials Today*, 36–40.

14. Scharber, M.C., Mühlbacher, D., Koppe, M., Denk, P., Waldauf, C., Heeger, A.J., Brabec, C.J. (2006). Design rules for donors in bulk-heterojunctiun solar cells — Towards 10% energy-conversion efficiency, *Adv. Mater.*, **18**: 789.

15. Cravino, A., (2007). Origin of the open circuit voltage of donor–acceptor solar cells : Do polaronic energy levels play a role?, *Appl. Phys Lett.*, **91**: 243502

16. Shaheen, S.E., Brabec, C.J., Sariciftci, N.S., Padinger, F., Fromherz, T., Hummelen, J.C., (2001). 2.5% efficient organic plastic solar cells , *Appl. Phys. Lett.*, **78**: 841–843.

17. Hoppe, H., Niggeman, M., Winder, C., Kraut, J., Hiesgen, R., Hinsch, A., Meissner, D., Sariciftci, N.S. (2004). Nanoscale morphology of conjugated polymer-fullerene-based bulk-heterojuntion solar cells, *Adv. Funct. Mater.*, **14**, 1005–1011.

18. Lee, J.K., Ma, W.L., Brabec, C.J., Yuen, J., Moon, J.S., Kim, J.Y., Lee, K., Bazan, G.C., Heeger, A.J. (2008). Processing additives for improved efficiency from bulk heterojunction solar cells, *J. Am. Chem. Soc.*, **130**, 3619–3623.

19. Li, G., Yao, Y., Yang, H., Shrotriya, V., Yang, G., Yang, Y., (2007). "Solvent annealing" effect in polymer solar cells based on poly(3-hexylthiophene) and methanofullerenes, *Adv. Funct. Mater.*, **17**, 1636–1644.

20. Hiorns, R., de Bettignies, R., Leroy, J., Bailly, S., Firon, M., Sentein, C., Khoukh, A., Preud'homme, H., Dagron-Lartigau, C., (2006). High molecular weights, polydispersities, and annealing temperatures in the optimization of bulk-heterojunction photovoltaic cells based on poly(3-hexylthiophene) or poly(3-butylthiophene). *Adv. Funct. Mater.*, **16**, 2263–2273.

21. Radbeh, R., Parbaile, E., Bouclé, J., Di Bin, C., Moliton, A., Coudert, V., Rossignol, F., Ratier, B., (2010). Nanoscale control of the network morphology of high efficiency polymer fullerene solar cells by the use of high material concentration in the liquid phase, *Nanotechnology*, **21**, 035201.

22. Baek, W.H. , Yang, H., Yoon, T.S., Kang, C.J., Lee, H.H., Kim, Y.S., (2009). Effect of P3HT:PCBM concentration in solvent on performances of organic solar cells, *Sol. Energy Mater. Sol. Cells*, **93**, 1263–1267.

23. Douhéret, O., Lutsen, L., Swinnen, A., Breselge, M., Vandewal, K., Goris, L., Manca, J. (2006). Nanoscale electrical characterization of organic photovoltaic blends by conductive atomic force microscopy, *Appl. Phys. Lett.*, **89**, 032107.

24. Coffey, D. C., Reid, O. G., Rodovsky, D. B., Bartholomew, G. P., Ginger, D. S. (2007). Mapping local photocurrents in polymer/fullerene solar cells with photoconductive atomic force microscopy, *Nano Lett.*, **7**, 738–744.

25. Leever, B. J., Durstock, M. F., Irwin, M. D., Hains, A. W., Marks, T. J., Pingree, L. S. C., Hersam, M. C. (2008). Spatially resolved photocurrent mapping of operating organic photovoltaic devices using atomic force photovoltaic microscopy, *Appl. Phys. Lett.*, **92**, 013302.

26. Nicholson, P. G., Castro, F. A. (2010). Organic photovoltaics: Principles and techniques for nanometre scale characterization, *Nanotechnology*, **21**, 492001.

27. Pingree, L. S. C., Reid, O. G., Ginger, D. S. (2009). Electrical scanning probe microscopy on active organic electronic devices, *Adv. Mater.*, **21**, 19–28.

28. Dante, M., Peet, J., Nguyen, T. Q. (2008). Nanoscale charge transport and internal structure of bulk heterojunction conjugated polymer/fullerene solar cells by scanning probe microscopy, *J. Phys. Chem. C*, **112**, 7241–7249.

29. Reid, O. G., Munechika, K., Ginger, D. S. (2008). Space charge limited current measurements on conjugated polymer films using conductive atomic force microscopy, *Nano Lett.*, **8**, 1602–1609.

30. Pingree, L. S. C., Reid, O. G., Ginger, D. S. (2009). Imaging the evolution of nanoscale photocurrent collection and transport networks during annealing of polythiophene/fullerene solar cells, *Nano Lett.*, **9**, 2946–2952.

31. Hamadani, B. H., Jung, S., Haney, P. M., Richter, L. J., Zhitenev, N. B. (2010). Origin of nanoscale variations in photoresponse of an organic solar cell, *Nano Lett.*, **10**, 1611–1617.

32. Dang, X. D., Tamayo, A. B., Seo, J., Hoven, C. V., Walker, B., Nguyen, T. Q. (2010). Nanostructure and optoelectronic characterization of small molecule bulk heterojunction solar cells by photoconductive atomic force microscopy, *Adv. Func. Mater.*, **20**, 3314–3321.

33. Rice, A. H., Giridharagopal, R., Zheng, S. X., Ohuchi, F. S., Ginger, D. S., Luscombe, C. K. (2011). Controlling vertical morphology within the active layer of organic photovoltaics using poly(3-hexylthiophene) nanowires and phenyl-C 61-butyric acid methyl ester, *ACS Nano*, **5**, 3132–3140.

34. Spadafora, E.J., Demadrille, R., Ratier, B., Grévin, B., (2010). Imaging the carrier photogeneration in nanoscale phase segregated organic heterojunctions by Kelvin probe force microscopy, *Nano Lett.*, **10** (9), 3337–3342.

35. Radbeh, R., Parbaile, E., Chakaroun, M., Ratier, B., Aldissi, M., Moliton, A., (2010). Enhanced efficiency of polymeric solar cells via alignment of carbon nanotubes, *Polym. Int.*, **59**, pp. 1514–1519.

Chapter 5

Theoretical Studies on Architectures of Straight Zigzag//Armchair Carbon Nanotube Junctions as Molecular Electronic Devices

Anurak Udomvecha[*] and Teerakiat Kerdcharoen[b]
[a]*Material Design and Discovery Research Laboratory, Department of Physics, Faculty of Science, Thaksin University (Songkhla Campus), Songkhla 90000, Thailand*
[b]*Center of Nanoscience and Nanotechnology and Department of Physics, Faculty of Science, Mahidol University, Rama 6 Road, Bangkok 10400, Thailand*
*nu_fermi1@yahoo.com

The intramolecular junctions of single-wall carbon nanotube (SWCNT), by various topological interfacial defects, have been theoretically investigated the geometric, electronic and rectifying properties. The interface geometries of junction were modeled by fusing the helical type varieties of zigzag and armchair tubes together via pentagon/heptagon (5/7) pair defects. The formation energy revealed an exothermic with the variation of nanotubes' diameter and chiralities. The electronic structures are dependent on the variation of tube's helicity and length. HOMO and LUMO

Advances in Nanodevices and Nanofabrication: Selected Publications from Symposium of Nanodevices and Nanofabrication in ICMAT2011
Edited by Qing Zhang, W. I. Milne, and Jianping Zou

ISBN 978-981-4364-54-6 (Hardcover), 978-981-4364-55-3 (eBook)
www.panstanford.com

states, particularly, exhibit an induction of even-odd "quantum size" oscillation since interfacial topologies have been changed through nanotubes' chiralities, like zigzag//zigzag and T-shapes junctions. The aspect of molecular rectification occurred on an even numbers of m in zigzag $(m,0)$ segments at interfacial junctions. Such helical tubes, in which the total molecular charge is zero, have a significant ability to increase the potential barrier across molecular junction due to the topological 5/7 defects around the tube circumference. The objective of this study is to extend the understanding of the inherent relationship between geometrical and electronic properties of intramolecular junctions for molecular electronic devices that are made up of carbon nanotubes.

5.1 Introduction

Currently carbon nanotubes [1] have been intensively researched and studied as a promising building block for developing various kinds of molecular-scale electronic in growth potential capability of computing devices and many kinds of applications [2]. With a highly perfect nanocylinder, they demonstrated a one-dimensional carbon-based molecule that can conduct electricity at room temperature. Moreover, many experimental and theoretical investigations have reported that the topological defect structures are possibly formed on the honeycomb lattice of typically carbon nanotubes [3,4]. Those defects can be used as fusing two or three different nanotubes together by covalent bond resulting in a tubular junction having different electronic properties. Therefore, an SWCNT can be either metallic or semi-conducting, depending on its chirality and diameter [5]. Hence, the combination of two segments of SWCNT having different helicity and diameter that connected by insertion of pentagon, heptagon rings and their mixed on the junction of hexagonal lattice of carbon nanotubes should handle the different electronic properties of carbon nanotubes.

Hence, we proposed the straight SWCNT intramolecular junctions (S-SWCNT-IMJ) [6] of the zigzag//armchair nanotubes via theoretical construction and examination. By joining of two nanotubes, various straight SWCNT-IMJ geometries were modeled to investigate the changes in electronic structures and their rectification manners upon the variation of tubes' diameter and their length. The formation energy is also presented to verifying the possible structural stability of SWCNT-IMJ structures.

5.2 Modeling and Computational Procedures

The combination of two carbon nanotube models with different chiralities can be done by inserting a number of pentagon, heptagon rings and/or a mixed of them into perfect cylinder of hexagonal lattice. This can govern the construction of the straight zigzag// armchair SWCNT junctions by joining the zigzag tubes with armchair tubes through a pentagon/heptagon (5/7) pair defects, as shown in Figure 5.1a,b. Two armchair tubules of (5,5) and (8,8) nanotubes were chosen to be fused with the varying zigzag (m,0), where m is 5 to 10 for joining with (5,5) and 6 to 10 for joining with (8,8) tubes, respectively. Both armchair tubes are intrinsically metallic, whereas the zigzag tubes are semi-conducting [7]. The orientation and increment of (5/7) pairs defects around the circumference depending on variation of tube's diameter is calculated. The notation of the straight junction nanotubes are denoted as "(m,0)//(5,5)" and "(m,0)//(8,8)". The extension of nanotube length in our model was defined by the unit cell layer from one to three layers, in which straight zigzag//armchair SWCNT junctions of three layers is shown in Figure 5.1.

The geometries of all SWCNT structures have been initiated by molecular mechanics (MM3) method. These structures were used as basis geometries for latter optimization by higher-level in quantum calculation techniques. The better optimized SWCNT-IMJ geometries and energetic of the entire structural variables were obtained from the semi-empirical AM1-RHF calculations. The SWCNT-IMJ formation energy, H_n, obtained from the semi-empirical AM1-RHF, is defined as

$$H_n = \frac{-E_{\text{SWCNT-IMJ}} + E_{(m,n)} + E_{(m,0)} + E_{\text{junc}}}{N}, \tag{5.1}$$

where $E_{\text{SWCNT-IMJ}}$ is the total energy of SWCNT-IMJ, $E_{(m,n)}$ and $E_{(m,0)}$ represent the total energy of each nanotube segment, and E_{junc} denotes the total energy of junction part (topological 5/7-pair defects structure). The more accurate electronic structure of all optimized SWCNT-IMJ structures was calculated by hybrid density functional theory of B3LYP and using all electron 6-31G(d) basis function. All computations have been done by using GAMESS package [8].

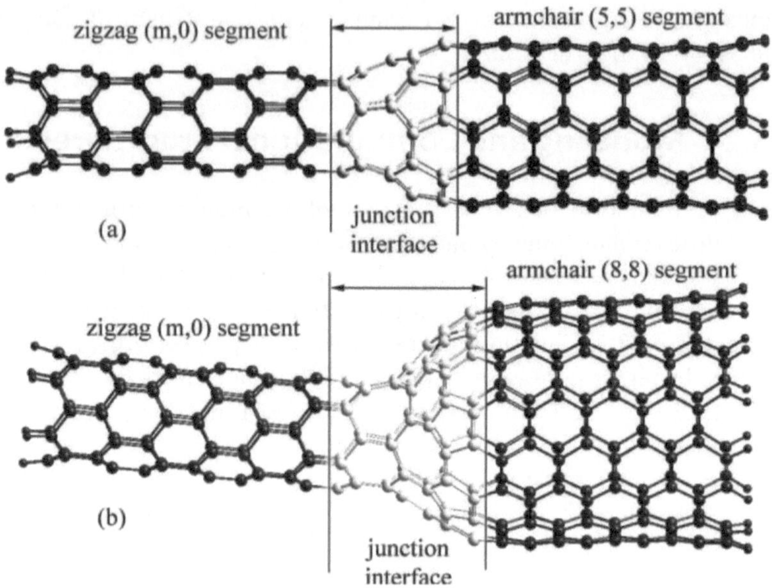

Figure 5.1 Straight-SWCNT-IMJ models represent the fusing of two carbon nanotubes: (a) $(m,0)//(5,5)$ and (b) $(m,0)//(8,8)$. The model shows the length parameter at 3 layers (or ~14 Å long) for both zigzag and armchair segments.

5.3 Results and Discussion

It has been reported, therefore, that topological 5/7 pair defects of SWCNT structures are less stable when compared with defect in which a hexagon is inserted between the pentagon and heptagon rings [9,10]. However, the significance of the obvious case of defect formation alignment should help pave the ways for the designation of molecular device architectures and be significant information for forthcoming synthesizing processes.

In this work, we computed the formation energy of both optimized SWCNT-IMJ structures obtained from the semi-empirical AM1 approach. The results performed that of such junction structures exhibited an endothermic reaction tendency which corresponds to less stable geometry (see Figure 5.2). However, the changing m indices of $(m,0)//(5,5)$ and $(m,0)//(8,8)$ junctions is dependent on the diameter size of the parent segments. That is therefore the fusing of large zigzag's diameter segments with armchair segments reveals the same decreasing trend of formation energy. The forming interfacial junctions of zigzag//armchair junctions gained lesser energies if diameter of zigzag segment reached the diameters of armchair segment. For instance, zigzag (9,0) (diameter ~7 Å) is equivalent with armchair (5,5) (diameter ~7 Å) while zigzag (10,0) (diameter ~8 Å) is equivalent with armchair (8,8) (diameter ~11 Å). In addition, the results unveils the combination of zigzag segments (in cases of *m* varied from 7 to 10) with diameter of (8,8) armchair segment gained more energies for forming interfacial junctions than smaller ones (i.e. (5,5) tube). For example, H_n of largest (10,0)//(5,5) structure is ~1000 kcal-mol^{-1}/C atom, whereas of (10,0)//(8,8) structure is ~1300 kcal-mol^{-1}/C atom. Except the joint of smallest diameter of zigzag segment (diameter ~3 - 4 Å) with armchair segments has remarkably made the junctions results in more stabilized straight junctions.

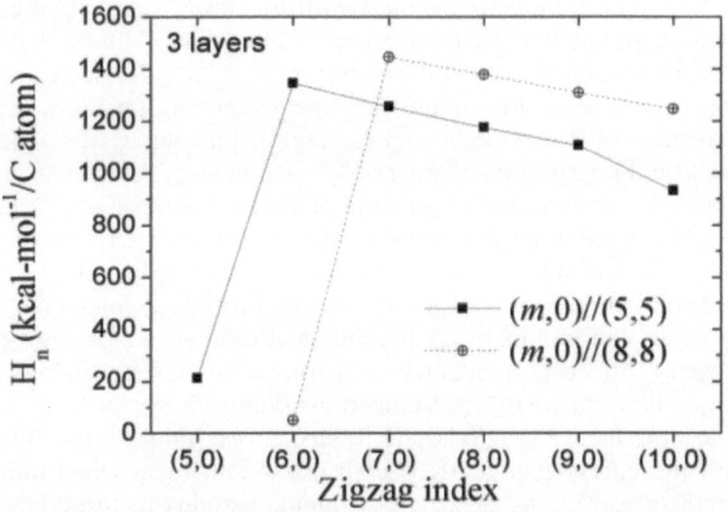

Figure 5.2 The formation energy (H_n) of the straight zigzag//armchair junctions with tubular length at three layers with respect to the tube's helicity.

These result are unlike the straight zigzag//zigzag structure types in ref. [11], in which the stability of (*m*,0)//(5,0) is exothermal process and independent on the diameter of the parent segments. Since the extra stability of straight zigzag//zigzag structures arises from increasing of number of 5/7 pairs defect. For straight zigzag//

armchair structures, however, are dependent on the diameter the parent segment and topological defect configuration and its orientation to form the junction.

The understanding molecular structure of CNT heterojunction is crucial choice for a central role of designation molecular electronic architecture through specifically topological 5/7 defects at interfacial junction. At interface, junction-variant of carbons bonding and different oriented configuration of 5/7 defects appointed to hold electronically controllable via indices of changing nanotube types and diameters. Hence, a database obtained from theoretical calculations is needed to assist in checking the possibility of making molecular electronic devices, e.g. molecular rectifier, from the SWCNT-IMJ structures.

On electronic features, the joint of zigzag and armchair nanotubes has been systematically studied. The calculations by the Kohn–Sham DFT frontier orbital reveal the even-odd "quantum size" oscillation via tube helical induction similar with the straight zigzag//zigzag and T-shape junctions in our previous works [11,12]. This oscillation is dependent on the orientations and number of 5/7 defects at the interfacial region. In comparison, the even-odd "quantum size" oscillation of the straight zigzag//zigzag junction is dependent solely on the associated number of 5/7 pair defects. For the zigzag// armchair junctions, the effect from associated number of 5/7 pair defects has induced the total molecular charge to switch between 0 and –1 for even and odd helical index numbers, respectively.

The energy gap (E_g) of zigzag//armchair junctions is demonstrated in Figure 5.3. It shows the difference in energy gap of zigzag//armchair junction results from tube's helicity and length parameters. For (m,0)//(5,5) junctions (Figure 5.3a), the change of helical tubule of zigzag (m,0)/(5,5) structures evidences the effects with the energy gap gently decreasing to small gap when tubule length of zigzag and armchair segments extends to three layers. Particularly, we found that the junction of linearly fusing between (9,0) and (5,5) nanotubes showed the smallest gap of about 0.15 eV at three layers. However, the variation of the armchair segment is resulted to the different in energy gap, as shown in Figure 5.3b. It was found that the helical changing of armchair segment is also play an importance role to energy gap of zigzag//armchair junctions. The average energy gap of straight (m,0)//(8,8) three-layer structures is less than 0.2 eV. The smallest energy gaps are reduced to *ca.* 0.1 eV for straight (6,0)//(8,8) three-layer junction. Because of the change of helical armchair segment makes thoroughly different topological configurations of 5/7 pair defect at the junction segment.

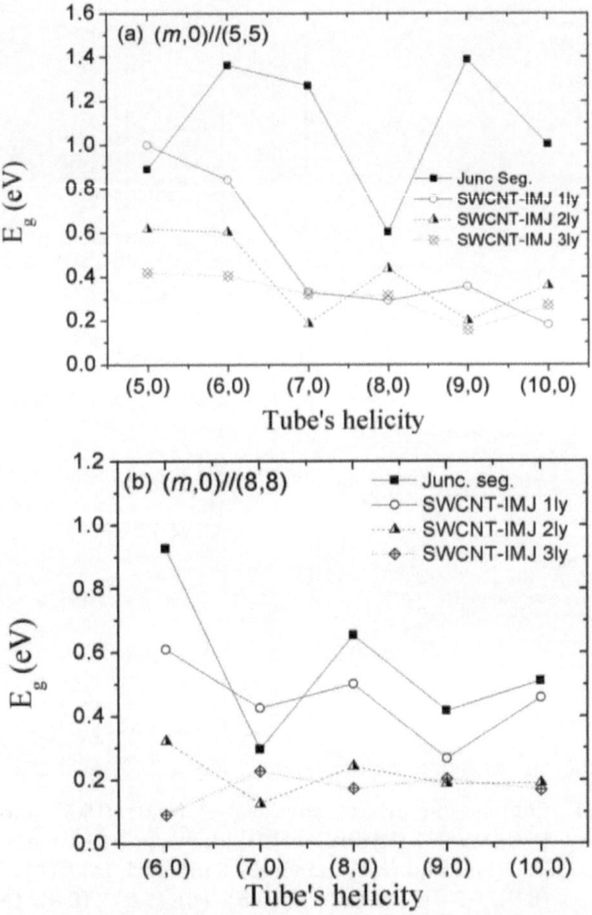

Figure 5.3 The DFT energy gap (E_g) of various zigzag//armchair SWCNT-IMJ structures plotted respect with the tube's helicity; (a) SWCNT-IMJ of $(m,0)//(5,5)$ and (b) SWCNT-IMJ of $(m,0)//(8,8)$.

However, only the description of frontier orbitals and energy gap from molecular orbital approximation methods are sufficiently unavailable representing the rectifying characteristic of SWCNT-IMJ. To identify the feasibility of zigzag//armchair junctions as molecular rectifier, we have to analyze the orbital energy levels of those IMJ structures compare with orbital energy of bare armchair and zigzag nanotubes, as shown in Figure 5.4.

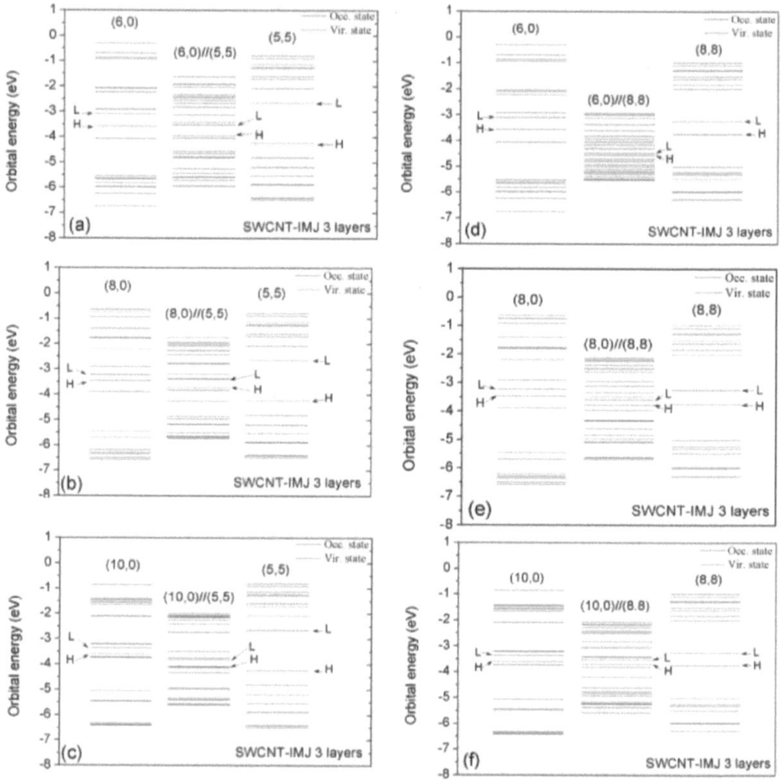

Figure 5.4 Comparison orbital energies of $(m,0)//(5,5)$ and $(m,0)//(8,8)$ SWCNT-IMJ structures with pristine $(m,0)$ and armchair (i.e. (5,5) and (8,8)) nanotubes, in which (a) (6,0)//(5,5), (b) (8,0)//(5,5), (c) (10,0)//(5,5), (d) (6,0)//(8,8), (e) (8,0)//(8,8), and (f) (10,0)// (8,8), respectively. Notice that upper most black-line of all three parts is frontier HOMO (H) state, and lowest red-line is frontier LUMO (L) state.

In general, the rectifying complexities of intramolecular junctions are composed of donor (D) and acceptor (A) subunits, which are connected via a spacer (σ) bridge. In case of zigzag// armchair models, the spacer region means a junction segment which one side is covalently bonding with zigzag $(m,0)$ segments through heptagon rings. On the opposite side, the spacer connected with armchair segments by the pentagon ring defects. The operation of flowing electron depends on the potential drop (ΔE_{LUMO}) across each subunit molecule is determined from the difference between E_{LUMO} and E_{LUMO+K} for an unoccupied state localized on the opposite (donor) side of molecule from the LUMO [13]. The calculated electronic charged on pentagon and heptagon member rings defects shown that the pentagon rings borrow π-electron density, while the heptagon rings donate π-electron density from surrounding C atoms [14]. In calculation, the rectification characteristic accessible has been seen on even *m* numbers of (*m*,0)//(5,5) and (*m*,0)//(8,8) three-layer junctions. It can be seen from Table 5.1, in which the energy difference of ΔE_{HOMO} and ΔE_{LUMO} between the donor (D)–spacer (σ) side and the spacer (σ)–acceptor (A) side is very close for (6,0)//(5,5), (8,0)//(5,5), (10,0)//(5,5), (6,0)//(8,8), (8,0)//(8,8), and (10,0)//(8,8) three-layer structures. As shown in Figure 5.4, the zigzag (*m*,0) segment increases in energy levels, while it is the opposite for the armchair segment. These can be attributed to the localization of HOMO and LUMO energy level on the donor and acceptor sides of the D-σ-A molecular complex, respectively. Hence the zigzag segment is donor side while the armchair segment is behaves like the acceptor side.

Table 5.1 Comparison between the energy difference, ΔE, of HOMO and LUMO between donor (D)–spacer (σ) side and spacer (σ)–acceptor (A) side

Variation of zigzag (*m*,0)	**(*m*,0)//(5,5)**				**(*m*,0)//(8,8)**			
	ΔE_{HOMO} (eV)		**ΔE_{LUMO} (eV)**		**ΔE_{HOMO} (eV)**		**ΔE_{LUMO} (eV)**	
	D–σ	σ–A	D–σ	σ–A	D–σ	σ–A	D–σ	σ–A
m = 5	1.29	1.79	1.10	0.63				
m = 6	0.37	0.30	0.43	0.88	0.99	0.81	1.36	1.22
m = 7	1.46	2.27	1.46	1.01	1.39	1.71	1.29	1.44
m = 8	0.25	0.52	0.18	0.75	0.31	0.02	0.38	0.35
m = 9	1.32	2.00	1.22	0.58	1.42	1.62	1.36	1.32
m = 10	0.47	0.15	0.45	1.16	0.09	0.05	0.17	0.28

To understand the process of electron tunneling through available orbital channels of these interfacial structures, the analysis of potential drop across molecular junction is necessary. As shown in Figure 5.4a,b,c, the HOMO energy of (*m*,0) and (*m*,0)//

(5,5) segments, in which m = 6, 8, 10, has equivalent energy levels that appear decoupling between donor and spacer moieties. On other site, the LUMO+4 energy of (m,0)//(5,5) segment of these junctions is equivalent with LUMO energy of (5,5) segment. There is confirmed decouple state on spacer and acceptor moieties also. For different kinds of armchair segment, the HOMO energy of the identical m numbers of (m,0)//(5,5) junctions occurs with (m,0)//(8,8) junctions. The equivalent energy levels of zigzag (m,0) and (m,0)//(8,8) segments resemble decoupling between donor and spacer moieties. For connection between (m,0)//(8,8) segment and (8,8) segment, it is found that the LUMO+2 level of (m,0)//(8,8) segment is equivalent to the LUMO level of (8,8) segment for the indices m of 8 and 10. Except (6,0)//(8,8) segment and (8,8) segment connecting, the LUMO+8 level is in place equivalent with LUMO level. That supported spacer and acceptor moieties are decoupling state.

Therefore in a rectifying mechanism, the acceptor moiety has a high dipole moment than donor moiety then the electron can transferred from acceptor moiety (1st excited state) to the donor moiety (ground state). If one can apply an appropriate bias voltage, the approximate potential drop (ΔE_{LUMO}) of three-part (m,0)//(5,5) molecular junctions by the difference of E_{LUMO}(donor)-E_{LUMO}(acceptor), which are for instance, 0.45 eV, 0.57 eV, and 0.71 eV for m = 6, 8, 10 of (m,0)//(5,5) junctions, respectively. While potential drop across the same even number of m for (m,0)//(8,8) junctions are obtain as 0.14 eV, 0.03 eV, and 0.11 eV. However, the characteristic of potential drop that across (6,0)//(8,8) and (8,0)//(8,8) junctions are performed behave abnormally decoupling at donor moiety in which the LUMO energy level of spacer moiety mismatch decouple with HOMO energy level of donor moiety. The resulting potential drop is increase in potential barrier. This is significant and can be attributed to the reason that the spacer moieties are typically aliphatic region with σ-bonds. Such bonds have hindrance for the tunneling of negatively charged electrons. A larger potential drop across a rectifier molecular junction is desirable to operate under a range of voltage in molecular circuit. So the variation of tube helicity of zigzag//armchair SWCNT-IMJ structures, where the total molecular charge is zero, has the ability to increase the potential barrier due to the topological defects around the tube circumference.

5.4 Conclusion

We have performed theoretical calculations to study the energetic and electronic properties as molecular rectifier of the straight zigzag//armchair SWCNT intramolecular junctions. We found that the formation energy revealed the endothermic reaction tendency with the variation of nanotubes' diameter and helicities. Except the (5,0)//(5,5) and (6,0)//(8,8) junctions have more stability than other zigzag//armchair junctions. The electronic properties of those junctions are dependent on the variation of tube's helicity and length. HOMO-LUMO levels exhibited even-odd "quantum size" oscillation from chiralities of nanotube, similar zigzag//zigzag and T-shape junctions. The energy gap indicates metallic or semi-conducting behavior depends on correlation between the tubes indices and their length. The aspect of molecular rectification has appeared in an even number of *m* of zigzag//armchair junctions. The variation of tube helicity of zigzag//armchair structures, where the total molecular charge is zero, has the ability to increase the potential barrier due to the topological defects around the tube circumference.

Acknowledgements

This work was financially supported by a Research Career Development for Southern Specification Region Grant (2009) to A.U. from the Commission of Higher Education.

References

1. Iijima, S. (1991). Helical microtubules of graphitic carbon, *Nature*, **354**, pp. 56–58.
2. Dekker, C. (1999). Carbon nanotubes as molecular wires, *Phys. Today*, **52**, pp. 22.
3. Ebbesen, T.W. and Takada, T. (1995). Topological and SP3 defect structures in nanotubes, *Carbon*, **33**, pp. 973–978.
4. Zhu, W., Rosen, A. and Bolton, K. (2008). Changes in single-walled carbon nanotube chirality during growth and regrowth, *J. Chem. Phys.*, **128**, pp. 124708.
5. Yao, Z., Postma, H.W.C., Balents, L. and Dekker, C. (1999). Carbon nanotube intramolecular junctions, *Nature*, **402**, pp. 273–276.

6. Meyyeppan, M. (2005). *Carbon Nanotubes Science and Applications* (Florida: CRC Press LLC).
7. An, K.H. and Lee, Y.H. (2006). Electronic-structure engineering of carbon nanotubes, *NANO: Brief Report and Reviews*, **1**, pp. 115–138.
8. Schmidt, M.W., Baldridge, K.K., Boatz, J.A., Elbert, S.T., Gordon, M.S., Jensen, J.H., Koseki, S., Matsunaga, N., Nguyen, K.A., Su, S.J., Windus, T.L., Dupuis, M. and Montgomery, J.A. (1993). The general atomic and molecular electronic structure system, *J. Comput. Chem.*, **14**, pp. 1347–1363.
9. Garau, C., Frontera, A., Quiñonero, D., Costa, A., Ballester, P. and Deyà, P.M. (2004). Structural and energetic features of single-walled carbon nanotube junctions: a theoretical *ab initio* study, *Chem. Phys.*, **303**, pp. 265.
10. Ye, Y., Zhang, M., Zhao, J., Liu, H. and Wang, N. (2008). *Ab initio* study on the structural, energetic and electronic features of the asymmetric armchair SWCNT junctions, *J. Mol. Struct. (THEOCHEM)*, **861**, pp. 79–84.
11. Shafiqzzaman, Md., Udomvech, A. and Kerdcharoen, T. (2007). Theoretical studies of pentagon-heptagon pair defects in carbon nanotube junctions, *Proceedings of 7th IEEE International Conference on Nanotechnology*, IEEE-NANO 2007, **2–5**, pp. 545–548.
12. Udomvech, A. and Kerdcharoen, T. (2008). Electronic structure of straight and T-shape singled-wall carbon nanotube junctions: A molecular quantum mechanics study, *Adv. Mater. Res.*, **55–57**, pp. 565–568.
13. Majumder, C., Mizaseki, H. and Kawazoe, Y. (2001) Molecular scale rectifer: Theoretical study, *J. Phys. Chem. A*, **105**, pp. 9454–9459.
14. Treboux, G., Lapstun, P. and Sliverbrook, K. (1999). An intrinsic carbon nanotube heterojunction diode, *J. Phys. Chem. B*, **103**, pp. 1871–1875.

Chapter 6

Energy Dissipation Analysis of Defected Carbon Nanotube Oscillators

Taiyu Guo,[a,*] Tony Weixi Ding,[a] Qing-Xiang Pei,[b] and Yong-Wei Zhang[b]
[a] *National University of Singapore High School of Mathematics and Science, 129957 Singapore*
[b] *Institute of High Performance Computing, 117528 Singapore*
*h0610040@highsch.nus.edu.sg

Molecular dynamics simulations are performed to study the energy dissipation of carbon nanotube oscillators with various defects. It is found that the Stone–Wales (SW) defects induce the largest energy dissipation, followed by the sp^3-hybridized defects in three rings and in one row. The sp^3-hybridized SW defects, however, result in the smallest energy dissipation among all the defected oscillators. These reductions in energy are central to the oscillatory instability of the nanotube oscillators. The energy dissipation mechanisms were discussed and changes in energy profiles were analyzed.

Advances in Nanodevices and Nanofabrication: Selected Publications from Symposium of Nanodevices and Nanofabrication in ICMAT2011
Edited by Qing Zhang, W. I. Milne, and Jianping Zou

ISBN 978-981-4364-54-6 (Hardcover), 978-981-4364-55-3 (eBook)
www.panstanford.com

6.1 Introduction

Carbon nanotubes (CNTs) have long attracted experimental and theoretical investigations due to their unique properties. Since the discovery of low friction multi-wall CNTs (MWNTs) [1, 2], studies on CNT-based gigahertz oscillators have been a focal point. These studies provide insights into the oscillatory behavior of nanotube oscillators. In our recent work, we analyzed the oscillatory behavior of CNT oscillators with sp^3-hybridized defects due to hydrogen adsorption [3]. It was found that oscillators with sp^3-hybridized defects in rows underwent larger oscillation decay compared to those containing the same number of defects in rings. Oscillators containing sp^3-hybridized SW defects were also analyzed. It was found that the presence of sp^3-hybridized defects can alleviate the inward distortions incurred by the SW defects and that certain configurations of which can almost completely remove such distortions.

In this paper, we discuss the energy dissipation mechanisms of various defected CNT oscillators, including those with sp^3-hybridized defects, SW defects, and sp^3-hybridized SW defects. Our results show that the extent of energy dissipation is strongly affected by the both the types and distributions of the defects, and that energy dissipation in the SW defected oscillator can be reduced by hydrogen adsorption.

6.2 Simulation Methodology

We have carried out molecular dynamics (MD) simulations with the reactive bond order (REBO) potential [4] as implemented in the software package LAMMPS and the Lennard-Jones pair potential. The combined form of the potentials can be represented by

$$E = \frac{1}{2}\sum\sum_{j \neq i}\left[E_{ij}^{REBO} + E_{ij}^{LJ}\right] \tag{6.1}$$

where E_{ij}^{REBO} describes short-ranged ($r < 2$ Å) C–C and C–H bonding interactions and E_{ij}^{LJ} adds longer-ranged van der Waals interactions with parameters ε_{CC} = 3.73 meV, σ_{CC} =3.4 Å and ε_{CH} = 0.65 meV, σ_{CH} = 2.59 Å for the C–C and C–H interactions, respectively [3]. The cutoff distance of the Lennard-Jones 12-6 potential was taken to be

12 Å. The REBO potential used here is essentially the same as the Brenner's potential [5], which has been widely used in studying the mechanical properties of CNTs and the oscillatory behavior of CNT oscillators.

The devised nanotube system comprises of an (10, 10) outer nanotube of 7.4 nm and an (5, 5) inner nanotube of 1.5 nm. The initial temperature is fixed at 1 K using the microcanonical ensemble and is allowed to rise due to frictional heating. The Verlet integration scheme is used with a time step of 0.2 fs and the inner and outer nanotubes are initially positioned to have a common axis.

6.3 Results and Discussion

6.3.1 Effect of sp^3-Hybridized Defects on Carbon Nanotube Oscillators

First, we investigate the cases where sp^3-hybridized defects are distributed in both rows and rings along the length of the CNT oscillators. Schematics of the oscillators are shown in Fig. 6.1b, and Fig. 6.1c, respectively. Compared with the defect-free oscillator as shown in Fig. 6.1a, the outer nanotube of the one-row defected oscillator (Fig 6.1b) is distorted into an almost tear-drop shape, where a single row of defected carbon atoms is shifted radially away from the center of the nanotube system. Also, it is observed that the presence of the ring defect induces local outward deformations only at the defected region, and that the overall shape of the outer nanotube is unchanged.

Distortion of the outer nanotube for both cases can be attributed to the hybridization states of the carbon orbitals. A transition from sp^2 to sp^3 state changes the bond angles from approximately 120° to 109.5°, which causes the defected carbon atoms to be further away from the center of the oscillator. As a result, cross-section of the oscillator is shifted away from its original circular shape.

The effect of embedding CNT oscillators with SW defect groups have already been studied previously [6]. It was found that when SW defects are introduced along the circumference of the outer nanotube, the radius of the defected region is decreased. The inner nanotube also experiences increased energy dissipation and intertube friction. In addition, the inner nanotube can only move through the defected region when it has enough kinetic energy to overcome the energy barrier associated with the defects. The inner nanotube can only oscillate between one end of the outer nanotube and the defected region when the peak kinetic energy of the axial modes of oscillation becomes sufficiently low. Based on these different motion states that can be achieved, a nano-switching mechanism was proposed [3, 7]. Simulations were performed for the above configuration where five SW defects were embedded onto a (10, 10) CNT oscillator. As seen in Fig. 6.1d, the region of defect can be clearly observed as a constriction in the nanotube.

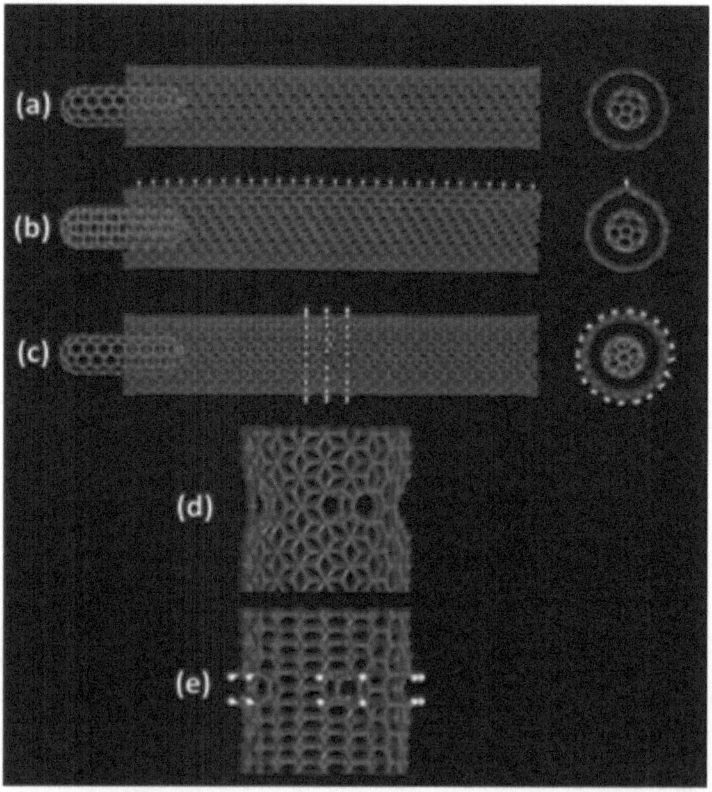

Figure 6.1 Schematics of the variations of carbon nanotube oscillator; (a) the defect-free oscillator, (b) the oscillator with sp^3-hybridized defects in one row, (c) the oscillator with sp^3-hybridized defects in three rings, (d) the oscillator with 5 SW defects placed around the outer nanotube, (e) the oscillator with sp^3-hybridized SW defects.

Referring to the displacement and combined kinetic and potential graph in Fig. 6.2, it is seen that the presence of the row defect significantly increases energy dissipation and the period of the intertube oscillations. The oscillators start off with roughly the same total energy of $E_i = -7.92$ eV. At the end of 13 complete cycles, it is observed that the corresponding changes in energy ΔE for both oscillators are $\Delta E_{\text{defect-free}} = 1.15$ eV and $\Delta E_{\text{1-row}} = 1.99$ eV, respectively. Comparing both ΔE values, it is clear that the one-row defect introduces to the system additional modes of energy dissipation. From the distorted shape of the defected oscillator, it can be evidently observed that the radius of the outer nanotube is smaller at the sites adjacent to the defected region. Previous studies by Liu *et al.* [8] have found that decreasing the diameter of an (10, 10) double-wall CNT oscillator increases the rate of energy dissipation. By introducing sp^3-hybridized defects to a (10, 10) defect-free oscillator, the inner nanotube becomes confined to a system of smaller radius, where energy dissipation is increased by the same mechanism.

In Fig. 6.2d, it is observed that the energy dissipation induced by the ring defect is comparatively larger than that by the row defected oscillator, with $\Delta E_{\text{3-rings}} = 3.52$ eV. This observation may be explained by the number of sp^3-hybridized defects present on each oscillator; the one-row defected oscillator contains 29 sp^3-hybridized carbon atoms, while the three-ring defect contains 60 such atoms. Therefore, the ring-defected oscillator experiences a relatively greater amount of distortion of the outer nanotube, which in turn increases energy dissipation during intertube oscillations.

The combined kinetic and potential energy profile in Fig. 6.2d shows that the SW defected oscillator experiences the greatest degree of energy dissipation among the five oscillators. In addition, the inner nanotube of the SW-defected oscillator undergoes half oscillation at $t = 270.1$ ps in Fig. 6.2a, where the peak kinetic energy of the inner nanotube decays to a point where there is insufficient energy to surmount the energy barrier. Simulations were also performed for the sp^3-hybridized SW defected oscillators. Surprisingly, it is observed in Fig 6.2d that energy dissipation is greatly reduced in the sp^3-hybridized oscillator, and that half oscillation no longer occurs. As the sp^3-hybridized outer nanotube restores its former defect-free shape, the oscillatory behavior that oscillator also more closely follows that of a defect-free oscillator.

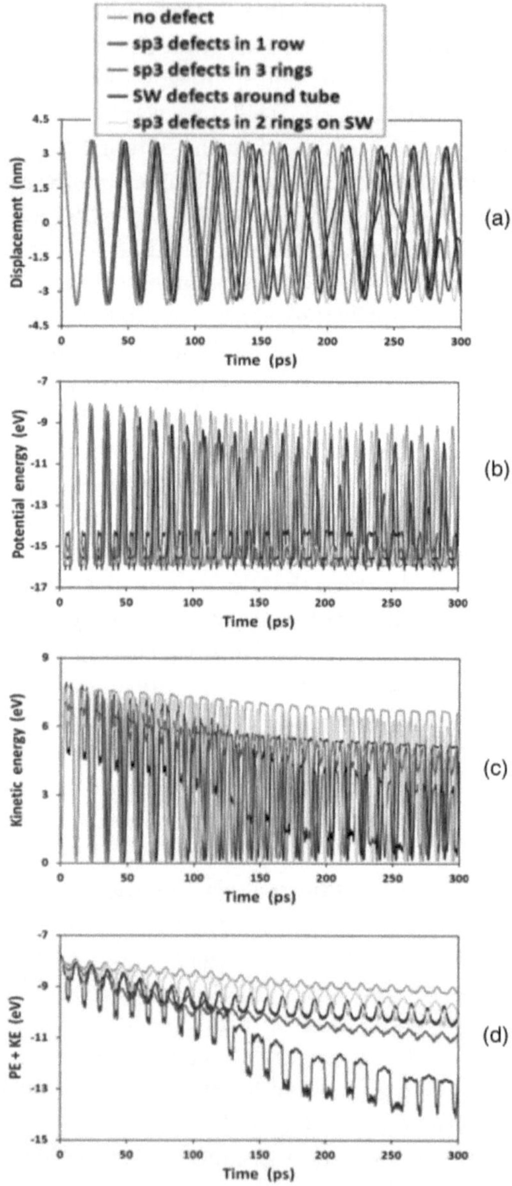

Figure 6.2 Oscillation profiles of the various oscillators. (a) The oscillation displacement over time, (b) the potential energy over time and (c) the kinetic energy over time and (d) the combined kinetic and potential energy profile of the inner nanotube.

6.3.2 Mechanisms of Energy Dissipation

Energy dissipation of the inner nanotube in a defect-free oscillator is attributed mainly to translational modes of intertube friction given by

$$E = \frac{1}{2}mv_0^2\,(1 - e^{(2ct/m)}) \tag{6.2}$$

In this equation, V_0 refers to the velocity of the inner nanotube along the axial direction, m denotes the mass of the inner nanotube and c is the coefficient of friction. When defects are incorporated into the outer nanotube, energy dissipation is seen to increase via additional modes of energy dissipation. Namely, they consist of rocking motion perpendicular to the axial direction and rotational motion of the inner nanotube [6–9]. Energy dissipation due to the rotational movements of the inner nanotube follows the equation:

$$E = \frac{1}{2}Iw^2\,(1 - e^{(2ct/I)}), \tag{6.3}$$

where I is the associated moment of inertia and w is the angular velocity of the inner nanotube. The presence of the additional modes of energy dissipation leads to greater decay in potential energy of the inner nanotube of the defected oscillators.

6.4 Conclusion

The energy dissipation mechanisms of CNT oscillators containing defects were studied using molecular dynamics simulations. It was found that the presence of sp^3-hybridized defects significantly increases energy dissipation of intertube oscillation. However, it was found that the adsorption of hydrogen atoms on the SW defects could reduce energy dissipation imposed by SW defect clusters. Our results may provide a defect-engineering approach to manipulate the energy dissipation, and thus the oscillatory behavior, of carbon nanotube oscillators.

References

1. Legoas, S. B., Coluci, V. R., Braga, S. F., Coura, P. Z., Dantas, S. O., Galvao, D. S. (2003). Molecular-dynamics simulations of carbon nanotubes as gigahertz oscillators, *Phys. Rev. Lett.*, **90**, 055504.

2. Zheng, Q., Jiang, Q., (2002). Multiwalled carbon nanotubes as gigahertz oscillators, *Phys. Rev. Lett.* **88**, 045503.
3. Guo, T.Y., Ding, T. W., Pei, Q.X., Zhang, Y.W., (2011). Effect of sp^3-hybridized defects on the oscillatory behavior of carbon nanotube oscillators, *Phys. Lett. A*, **375**, 2400–2404.
4. Stuart, S.J., Tutein, A.B., Harrison, J.A., (2000). A reactive potential for hydrocarbons with intermolecular interactions, *J. Chem. Phys.*, **112**, 6472–6486.
5. Brenner, D.W., Shenderova, O.A., Harrison, J.A., Stuart, S.J., Ni, B., Sinnott, S.B., (2002). A second-generation reactive empirical bond order (REBO) potential energy expression for hydrocarbons, *J. Phys. Condens. Matter*, **14**, 783–802.
6. Liu, P., Gao, H. J., Zhang, Y. W., (2008). Effect of defects on oscillation characteristics and instability of carbon nanotube-based oscillators, *Appl. Phys. Lett.*, **93**, 083107.
7. Liu, P., Zhang, Y. W. (2010). Motion control in double-walled carbon nanotube systems using a Stone–Thrower–Wales defect cluster, *J. Phys. D Appl. Phys.*, **43**, 445404.
8. Liu, P., Zhang, Y. W., Lu, C., (2006). Analysis of the oscillatory behavior of double-walled carbon nanotube oscillators, *Carbon*, **44**, 27–36.
9. Zhao, Y., Ma, C.C., Chen, G., Jiang, Q., (2003). Energy dissipation mechanisms in carbon nanotube oscillators, *Phys. Rev. Lett.*, **91**, 175504.

Chapter 7

Single-Wall Carbon Nanotube-Based Transparent and Conductive Films

Zhiyu Yan,[a] Xuena Luan,[b] Yu Fan,[a] and Qing Zhang[a,*]
[a] *Microelectronics Centre, School of Electrical and Electronic Engineering, Nanyang Technological University, 639798 Singapore*
[b] *School of Chemical and Bio-medical Engineering, Nanyang Technological University, 637459 Singapore*
*eqzhang@ntu.edu.sg

In this chapter, the performances of the single-wall carbon nanotube (SWCNT) based transparent and conductive films (TCFs) are investigated through optimizing several key fabrication conditions. We find that, with the SWCNT suspension centrifuged at 10000 rpm and glass substrate treated with APTES, soft-baking at 105°C, hard-baking at 300°C, HNO_3 acid and PEDOT post treatments efficiently improve the conductivity and transmittance of the SWCNT-based TCFs.

7.1 Introduction

Transparent and conductive films (TCFs) have been adopted as essential components of liquid crystal displays, touch screens, solar

---

Advances in Nanodevices and Nanofabrication: Selected Publications from Symposium of Nanodevices and Nanofabrication in ICMAT2011
Edited by Qing Zhang, W. I. Milne, and Jianping Zou

ISBN 978-981-4364-54-6 (Hardcover), 978-981-4364-55-3 (eBook)
www.panstanford.com

cells, etc. For commercial applications, most of TCFs are indium tin oxide (ITO) films, which are brittle and become more expensive as the supply of indium runs low. In contrast, SWCNTs have been regarded as one potential candidate to replace ITO [1]. A typical SWCNT has a diameter between 1.0 and 2.0 nm, and its resistance can be as low as the quantum resistance of 14 KΩ. Thus, an SWCNT network with an appropriate number density can possess high electrical conductivity and high transmittance in the visible range. For commercial use, the minimum optical and electrical requirements for TCFs are the transmittance $T > 90\%$ and sheet resistance $R_s < 100$ Ω/sq, respectively [2]. That is to say that the commercially useful TCFs should have higher transmittance and lower sheet resistance. Simply increasing SWCNT concentration in an SWCNT-based TCF could definitely increase the conductance of the film. However, the more SWCNTs are involved in the film, the less transparent the film is. Thus, the main challenge of the SWCNT-based TCFs is to tackle this trade-off problem between the transmittance and sheet resistance.

The main processes of fabricating the SWCNT-based TCFs include SWCNT purification, dispersion, deposition and functionalization. The purpose of purification is to remove impurities produced during the synthesis of SWCNTs, including metallic catalyst particles, graphite, amorphous carbon, and carbon nanoparticles. Physical separation, gas or vapor phase oxidation and liquid phase oxidation are the common ways to purify SWCNTs. Centrifugation and filtration are often used physical separation techniques to remove those large mass density or/and oversized particles. Gas or vapor phase oxidation attacks metallic impurities by oxidizing their carbonaceous shells to make them easy to be dissolved by acid treatments [3]. Liquid phase oxidation is often carried out in acid solutions of strong oxidants such as HCl, HNO_3, H_2SO_4, and HF to dissolve metallic catalytic particles and amorphous carbon [4–6]. These processes are found to strongly affect the performances of final SWCNT-based TCFs. In addition, the processing time duration, concentration of chemicals, anneal temperatures, etc., must be well controlled to avoid damaging SWCNTs [7,8].

Dispersion separates SWCNTs from each other and prevents them from agglomeration by van der Waals attraction in liquid solvents. Ultrasonication is a common mechanical method that causes deagglomeration and milling of micrometer- and nanometer-size materials because of the effect of cavitation. Special chemical treatments during dispersion can functionalize the surface of SWCNTs and change their wetting characteristics [9,10]. Surfactants are used in dispersion to promote the separation of SWCNTs. The interaction between SWCNTs and surfactants plays a crucial role and the concentration of surfactants used should be just appropriate in order to optimize their dispersing power [11,12].

Main methods for depositing SWCNTs onto a substrate include spin coating, spray coating, dip coating, rod coating, layer by layer, electrophoretic deposition, chemical vapor deposition, etc. The smoothness and thickness of the SWCNT-based TCFs determine the conductivity and transmittance and they can be controlled through adjusting the processing conditions [13–17].

Functionalization improves the solubility and conductivity of SWCNTs by adding functional groups onto the SWCNT surface though chemical synthesis methods. Non-covalent approach includes polymer wrapping, biomolecule binding, metal ion binding, etc. [18]. While covalent approach includes oxidizing SWCNTs in concentrated HNO_3 or the mixture of H_2SO_4 and HNO_3 to introduce carboxyl group (–COOH), treating SWCNTs with KOH to introduce hydroxyl groups (–OH), reacting with HF to introduce fluoro group (–F), etc. [19,20]. Another method is chemical doping which could decrease the sheet resistance at tube–tube junctions efficiently. These chemicals include graphene, Au, conductive polymers, etc. [21–23]. However, it should be pointed out that covalent functionalization typically causes a significant decrease in the conductivity of the functionalized SWCNTs as covalent functional groups change the bond structures of SWCNTs [24].

In this chapter, we will focus on the influences of key fabrication process conditions on the performances of the SWCNT-based TCFs. The number of spin-coatings, centrifugation, (3-aminopropyl) triethoxysilane (APTES) treatment, soft-baking, hard-baking, acid treatment, and poly (3,4-ethylenedioxythiophene) (PEDOT) treatment are analyzed in details.

7.2 Experimental Procedure

In our experiments, SWCNTs are dispersed in deionized water with 0.555 g/L. Triton X-100 is added into this SWCNT suspension as the surfactant with 1 wt%. The suspension is ultrasonicated for 2 h and centrifuged at 10000 rpm for 1 h to well disperse SWCNTs and remove large size impurity particles. The glass substrate is cleaned by ultrasonication in acetone and isopropyl alcohol solution for 10 min, respectively. It is immersed into piranha solution (H_2SO_4: H_2O_2 = 3:1) for 20~30 min and then into isopropyl alcohol solution with 0.3 wt% APTES for 1 h at 60°C to make the glass substrate surface hydrophilic. SWCNT suspension is spin-coated onto the glass substrate. The rotation speed and time duration are optimized at 2800 rpm and 30 s, respectively. The SWCNT coatings are soft-baked at 105°C for 3~5 min at the interval between two successive spin-coatings to vaporize the moisture left on the surface. Upon the completion of SWCNT deposition, the SWCNT-based TCFs are hard-baked at 300°C for at least 15 min in the ambient environment to decompose Triton X-100. After appropriate acid and PEDOT treatment, the fabricated SWCNT-based TCFs are characterized with UV-vis spectrophotometer, four-point probe and SEM, respectively.

7.3 Results and Discussion

7.3.1 The Effects of Number of Spin Coatings

Figures 7.1(a) and (b) show the SEM images of two SWCNT-based TCFs made from centrifuged SWCNT suspension with (a) 20 and (b) 40 spin-coatings, respectively. The transmittance is 98% at 550 nm for the TCF in Fig. 7.1(a) and 95% for the TCF in Fig. 7.1(b). From the sheet resistances given in Table 7.1, one can see that the TCF with 40 spin-coatings has a denser SWCNT network, leading to a large conductivity and a low transmittance in comparison with the TCF with 20 spin-coatings.

7.3.2 The Effects of Centrifugation

Figures 7.1(c) and (d) show the SEM images of two TCFs made from non-centrifuged SWCNT suspension with (c) 20 and (d) 40 spin-coatings, respectively. From Table 7.1, one can find that the TCFs made from centrifuged SWCNT suspension have smaller sheet resistance and variation than those in Figs. 7.1(c) and (d). The reason is that the TCFs made from the centrifuged SWCNT suspension form a smoother SWCNT network without involving large-size particles such as large bundles of aggregated SWCNTs, large-size graphite and carbon nanoparticles, etc. These networks smoothen electron transport, leading to the improvement of conductivity. Absence of these large particles also helps to form uniform TCFs otherwise large particles deposited on the substrate would block SWCNTs from depositing on the locations where they are, resulting in a large variation of sheet resistances.

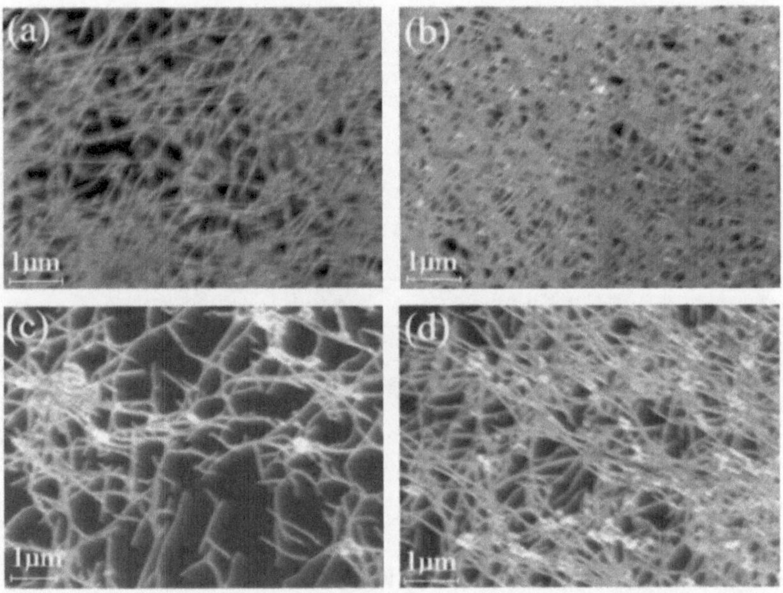

Figure 7.1 SEM images of the TCFs made from centrifuged SWCNT suspension with (a) 20 spin-coatings, (b) 40 spin-coatings. The TCFs made from non-centrifuged SWCNT suspension with (c) 20 spin-coatings, (d) 40 spin-coatings.

7.3.3 The Effects of APTES Treatment

Figure 7.2 compares the sheet resistance of the SWCNT-based TCFs on the glass substrates with or without APTES treatments. With the same spin-coatings of identical SWCNT suspension, the TCF on the APTES treated substrate has a smaller sheet resistance than without APTES treatment. Note that, in the SWCNT suspension, –OH groups of Triton X-100 wrap SWCNTs. This leaves hydrophilic –OH groups exposed. When APTES interlayer is formed on the surface of glass, $-NH_2$ group of APTES reacts with –OH group of Triton X-100; so hydrogen bonding forms. If SDS is used as a surfactant (in Sections 7.3.7 and 7.3.8), APTES can still work through electrostatic attraction between the negatively charged head of SDS and the positively charged $-NH_2$ groups of APTES [25]. Thus, APTES treated glass substrate attracts more SWCNTs, forming a denser SWCNT network and resulting in a low sheet resistance.

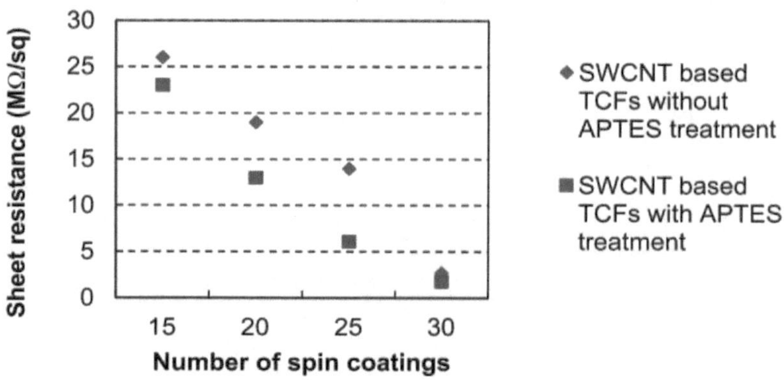

Figure 7.2 Sheet resistances of the SWCNT-based TCFs on the glass substrates with (squares) or without (diamonds) APTES treatment, respectively.

7.3.4 The Effects of Soft Baking

Soft-baking is introduced in between two successive spin-coatings of SWCNT suspension. Table 7.2 shows the effects of the soft-baking. With the same spin-coatings, the TCFs made with soft-baking have smaller sheet resistances and variations than those without soft-baking. Soft-baking removes the moisture left by spin-coating, dries the deposited SWCNT network and enables a strong adhesion of SWCNT coating onto the substrate prior to the next round of spin-coating. Thus, deposited SWCNT density increases with the number of spin-coatings, rather than saturates after a certain number of spin-coatings, resulting in an increase in the conductance of the SWCNT-based TCFs.

Table 7.1 Sheet resistances of the SWCNT-based TCFs in Figs. 7.1(a–d)

SWCNT-based TCFs	Position 1 (MΩ/sq)	Position 2 (MΩ/sq)	Position 3 (MΩ/sq)	Position 4 (MΩ/sq)	Position 5 (MΩ/sq)	Average (MΩ/sq)	Variance
Fig. 7.1(a)	1.0	0.85	1.0	0.96	1.1	0.98	0.00812
Fig. 7.1(b)	0.41	0.42	0.58	0.34	0.31	0.41	0.0110
Fig. 7.1(c)	3.3	3.8	4.0	3.2	4.0	3.7	0.148
Fig. 7.1(d)	0.42	0.56	0.58	0.5	0.86	0.58	0.0277

Table 7.2 Sheet resistances of the SWCNT-based TCFs with and without soft-baking

SWCNT-based TCFs	Position 1 (MΩ/sq)	Position 2 (MΩ/sq)	Position 3 (MΩ/sq)	Position 4 (MΩ/sq)	Position 5 (MΩ/sq)	Average (MΩ/sq)	Variance
Without soft-baking	1.0	1.5	1.1	1.4	1.6	1.3	0.067
With soft-baking	0.26	0.26	0.25	0.27	0.21	0.25	0.00055

7.3.5 The Effects of Hard Baking

Figure 7.3 shows the influences of hard-baking on the sheet resistance of the SWCNT-based TCFs. Triton X-100, the surfactant used in dispersing SWCNTs, is decomposed quickly at 300°C. From Fig. 7.3, one can see that the sheet resistance reduces significantly after 15 min hard-baking. Long time hard-baking does not further reduce the sheet resistance of the SWCNT-based TCFs. This suggests that 15 min hard-baking is sufficient to remove Triton X-100 residues in the deposited SWCNT network.

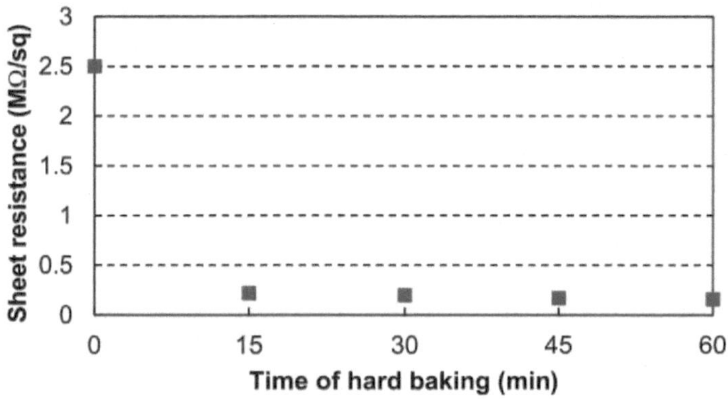

Figure 7.3 Sheet resistances of the SWCNT-based TCFs before and after hard-baking treatments.

7.3.6 The Effects of Acid Treatment

After hard-baking, the SWCNT-based TCFs are immersed into 65% HNO_3 for 20s, rinsed with deionized water, and soft-baked for 10 min to enhance adhesion between SWCNTs and the glass substrates. It is found that the sheet resistance is reduced by 61.9% on average, as shown in Fig. 7.4.

Decrease in the resistance up on the acid treatment could result from two factors. A smoother SWCNT network can be obtained up on the acid treatment because 65% HNO_3 dissolves most of the impurities which could be deposited on the substrate during spin-coating process and block electron transport. In addition, during the acid treatment, functional groups, such as –COOH, could attach to the sidewall of SWCNTs. The functional groups are conductive and help electrons flow from one SWCNT to another.

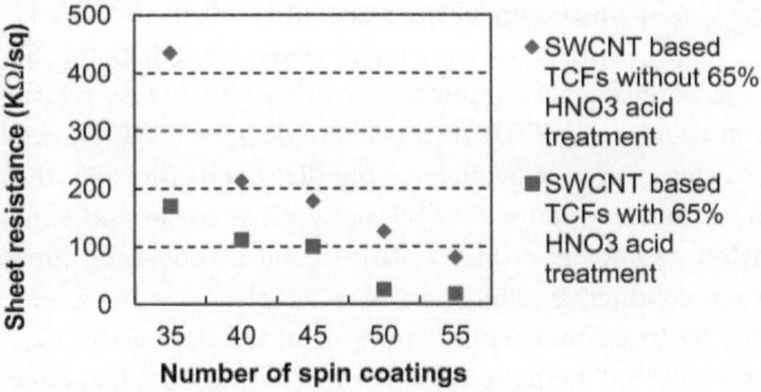

Figure 7.4 Sheet resistances of the SWCNT-based TCFs before (diamonds) and after (squares) 65% HNO_3 acid treatment.

7.3.7 The Effects of PEDOT Treatment

PEDOT is a conductive polymer. Incorporation of PEDOT into a SWCNT network could enhance its conductance [26]. 1 wt% SDS is used as the surfactant in the following experiments. The SWCNT suspension is mixed with PEDOT. The mixture is deposited onto the APTES treated glass substrate. As PEDOT vaporizes at 300°C, hard-baking process is not implemented. Using SDS as the surfactant, rather than Triton X-100, is because SDS can be removed by rinsing with deionized water at the ambient temperature. As illustrated in Fig. 7.5, 2 wt% PEDOT causes a significant improvement in the conductivity of the TCFs.

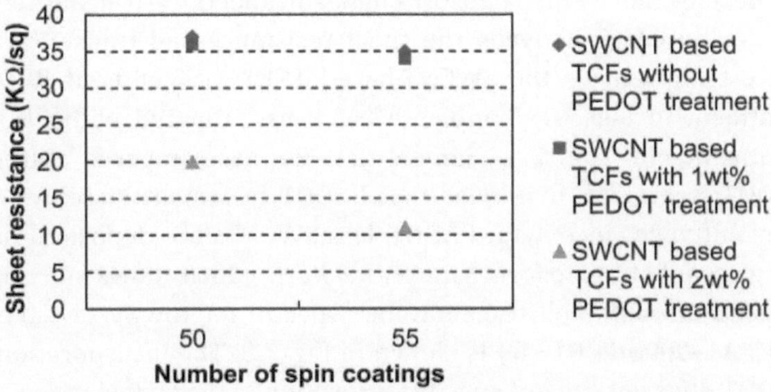

Figure 7.5 Sheet resistances of the SWCNT-based TCFs without PEDOT treatments (diamonds), with 1 wt% PEDOT treatment (squares) and with 2 wt% PEDOT treatment (triangles).

The conductance improvement could result from three factors. In Fig. 7.6(a), PEDOT fills up most voids existed in the SWCNT network, resulting in a smooth network which is less resistant to electron transport. PEDOT also helps to disperse SWCNTs well and reduces aggregation to avoid large bundles depositing onto the glass substrate. Besides, pure SWCNT network is loose and randomly entangled as shown in Fig. 7.6(b). In these tube–tube junctions, some are conductive, which are easy for electrons to tunnel from one SWCNT to another, while some are not and difficult for electrons to tunnel. PEDOT helps to form better electrical contacts between SWCNTs, which decreases the tunneling barriers in tube–tube junctions and improves the intertube electron transfer.

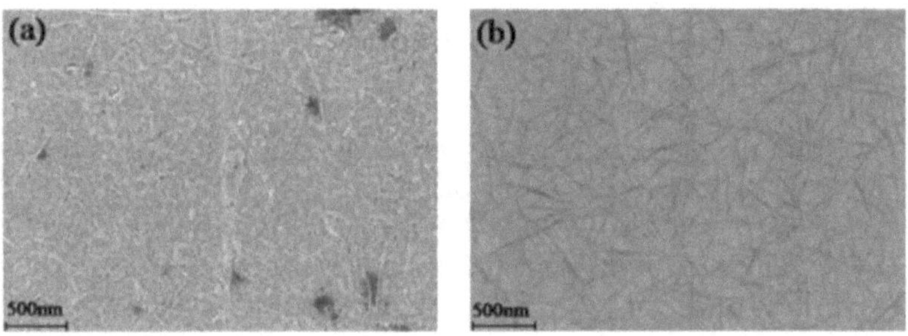

Figure 7.6 SEM images of (a) SWCNT-based TCF with 2wt% PEDOT treatment (surfactant: SDS), (b) SWCNT-based TCF without PEDOT treatment (surfactant: SDS).

Besides, since PEDOT absorbs light and affects the transmittance, the relationships between the sheet resistance and transmittance are established for the SWCNT-based TCFs with/without PEDOT treatment in Fig. 7.7. Samples A1~A4 are prepared with PEDOT treatment. In contrast, Sample B1~B4 are prepared without PEDOT treatment. It is clear that PEDOT treatment increases the transmittance since, with PEDOT, less SWCNTs are deposited onto the glass substrate to form a looser network, which allows more light passing through. Full transmittance spectra for the SWCNT-based TCFs A1~A4 and B1~B4 is shown in Fig. 7.8. The incorporation of PEDOT does not cause a specific absorption in the visible range.

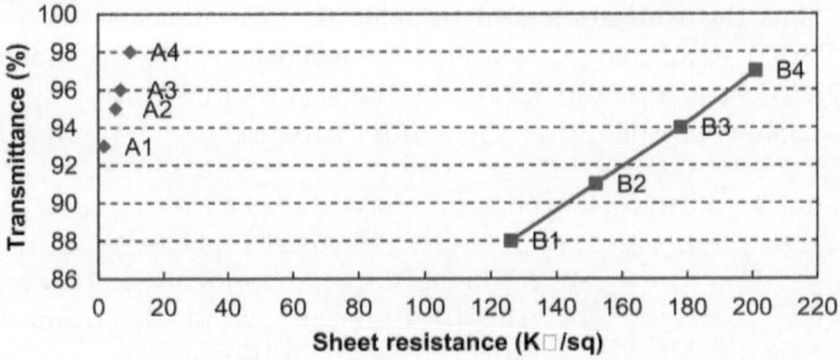

Figure 7.7 The sheet resistance vs transmittance at 550 nm wavelength for the SWCNT-based TCFs with (diamonds) and without (squares) PEDOT treatment.

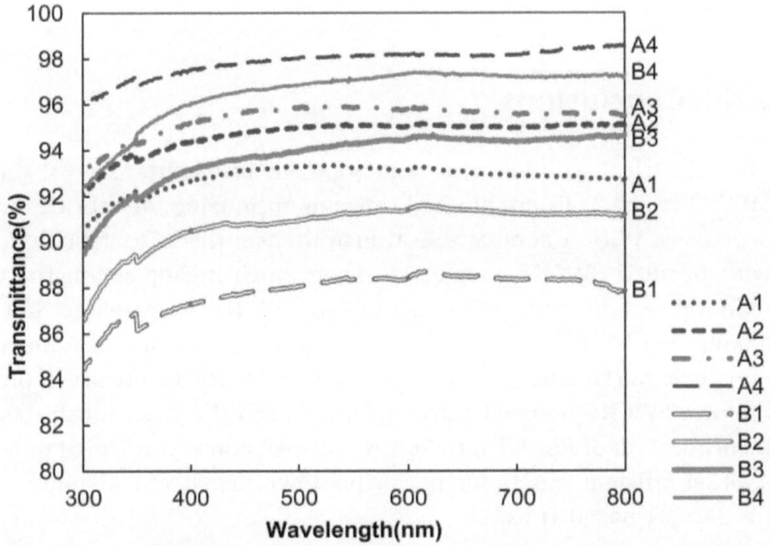

Figure 7.8 Transmittance spectra for the SWCNT-based TCFs with and without PEDOT treatment. Samples A1~A4 and B1~B4 are the same samples as in Fig. 7.7, correspondingly.

7.3.8 The Effects of the Combination of Acid and PEDOT Treatment

The SWCNT-based TCFs with PEDOT treatment are further treated with acid by immersing them into 65% HNO_3 for 20 s. It is found that the sheet resistance is reduced apparently (Table 7.3).

Since the oxidization of HNO_3 delocalizes the atoms on the chain of PEDOT, extra charges as well as the electron density are induced [27]. This could be the origin that the acid treatment can improve the conductivity of PEDOT and SWCNT network.

Table 7.3 Effects of the combination of acid and PEDOT treatment

	SWCNT-based TCFs with 1wt% PEDOT treatment (KΩ/sq)		SWCNT-based TCFs with 2wt% PEDOT treatment (KΩ/sq)	
Before 65% HNO_3 acid treatment	34	36	11	20
After 65% HNO_3 acid treatment	6.8	10	1.7	5.3

7.4 Conclusions

The electrical conductivity and optical transmittance of the SWCNT-based TCFs are studied through optimizing the fabrication processes. With a comparable transmittance, the TCFs fabricated with purified SWCNT suspension have much higher conductivity than those fabricated with unpurified SWCNT suspension. Soft-baking, hard-baking and acid treatment are necessary to remove moisture, surfactant residues and nanoparticles to ensure more uniform SWCNT network strongly adhered to the glass substrates. Incorporation of PEDOT into SWCNT suspension is confirmed to be another efficient way to further bring down the sheet resistance of the SWCNT-based TCFs.

References

1. Feng, C., Liu, K., Wu, J. S., Liu, L., Cheng, J. S., Zhang, Y., Sun, Y., Li, Q., Fan, S., and Jiang, K. (2010). Flexible, stretchable, transparent conducting films made from superaligned carbon nanotubes, *Adv. Funct. Mater.*, 20, 885–891.

2. Scardaci, V., Coull, R., and Coleman, J. N. (2010). Very thin transparent, conductive carbon nanotube films on flexible substrate. *Appl. Phys. Lett.*, 97, 023114.

3. Montoro, L. A., and Rosolen, J. M. (2006). A multi-step treatment to effective purification of single-walled carbon nanotubes, *Carbon,* 44, 3293–3301.

4. Reyhania, A., Mortazavi, S. Z., Golikand, A. N., Moshfegh, A. Z., and Mirershadi, S. (2008). The effect of various acids treatment on the purification and electrochemical hydrogen storage of multi-walled carbon nanotubes, *J. Power Sources,* 183, 539–543.

5. Salernitano, E., Giorgi, L., Makris, Th. D., Giorgi, R., Lisi, N., Contini, V., and Falconieri, M. (2007). Purification of MWCNTs grown on a nanosized unsupported Fe-based powder catalyst, *Diamond Related Mater.,* 16, 1565–1570.

6. Mathur, R. B., Seth, S., Lal, C., Rao, R., Singh, B. P., Dhami, T. L., and Rao, A. M. (2007). Co-synthesis, purification and characterization of single- and multi-walled carbon nanotubes using the electric arc method, *Carbon,* 45, 132–140.

7. Ko, F. H., Lee, C. Y., Ko, C. J., and Chu, T. C. (2005). Purification of multi-walled carbon nanotubes through microwave heating of nitric acid in a closed vessel, *Carbon,* 43, 727–733.

8. Schonfelder, R., Rummeli, M. H., Gruner, W., Loffler, M., Acker, J., Hoffmann, V., Gemming, T., Buchner, B., and Pichler, T. (2007). Purification-induced sidewall functionalization of magnetically pure single-walled carbon nanotubes, *Nanotechnology,* 18, 375601–375608.

9. Ma, P. C., Siddiqui, N. A., Marom, G., and Kim, J. K. (2010). Dispersion of Carbon Nanotubes in Liquids, *Composites: Part A,* 41, 1345–1367.

10. Osorio, A. G., Silveira, I. C. L., Bueno, V. L., and Bergmann, C. P. (2008). $H_2SO_4/HNO_3/HCl$—Functionalization and its effect on dispersion of carbon nanotubes in aqueous media, *Appl. Surf. Sci.,* 255, 2485–2489.

11. Rastogi, R., Kaushal, R., Tripathi, S. K., Sharma, A. L., Kaur, I., and Bharadwaj, L. M. (2008). Comparative study of carbon nanotube dispersion using surfactants, *J. Colloid Interface Sci.,* 328, 421–428.

12. Bystrzejewski, M., Huczko, A., Lange, H., Gemming, T., Buchner, B., and Rummeli, M. H. (2010). Dispersion and diameter separation of multi-wall carbon nanotubes in aqueous solutions, *J. Colloid Interface Sci.,* 345, 138–142.

13. de Andrade, M. J., Lima, M. D., Skákalová1, V., Bergmann, C. P., and Roth, S. (2007). Electrical properties of transparent carbon nanotube networks prepared through different techniques, *Phys. Status Solid Rapid Res. Lett.,* 1, 178–180.

14. Paul, S., Kang, Y. S., Yim, J. H., Cho, K. Y., and Kim, D. W. (2010). Effect of surfactant and coating method on the electrical and optical properties

of thin conductive films prepared with single-walled carbon nanotubes, *Curr. Appl. Phys.*, 10, E101–E104.

15. Dan, B., Irvin, G. C., and Pasquali, M. (2009). Continuous and scalable fabrication of transparent conducting carbon nanotube films, *ACS Nano*, 3, 835–843.
16. Yu, X., Rajamani, R., Stelson, K. A., and Cui, T. (2008). Fabrication of carbon nanotube based transparent conductive thin films using layer-by-layer technology, *Surf. Coatings Technol.*, 202, 2002–2007.
17. Lima, M. D., de Andrade, M. J., Bergmann, C. P., and Roth, S. (2008). Thin, conductive, carbon nanotube networks over transparent substrates by electrophoretic deposition, *J. Mater. Chem.*, 18, 776–779.
18. Porro, S., Musso, S., Vinante, M., Vanzetti, L., Anderle, M., Trotta, F., and Tagliaferro, A. (2007). Purification of carbon nanotubes grown by thermal CVD, *Physica E*, 37, 58–61.
19. Kukovecz, A., Kramberger, Ch., Holzinger, M., Kuzmany, H., Schalko, J., Mannsberger, M., and Hirsch, A. (2002). On the stacking behavior of functionalized single-wall carbon nanotubes, *J. Phys. Chem. B*, 106, 6374–6380.
20. Kelly, K. F., Chiang, I. W., Mickelson, E. T., Hauge, R. H., Margrave, J. L., Wang, X., Scuseria, G. E., Radloff, C., and Halas, N. J. (1999). Insight into the mechanism of sidewall functionalization of single-walled nanotubes: An STM study, *Chem. Phys. Lett.*, 313, 445–450.
21. King, P. J., Khan, U., Lotya, M., De, S., and Coleman, J. N. (2010). Improvement of transparent conducting nanotube films by addition of small quantities of graphene, *ACS Nano*, 4, 4238–4246.
22. Yang, S. B., Kong, B. S., Kim, D. W., Baek, Y. K., and Jung, H.T. (2010). Effect of Au doping and defects on the conductivity of single-walled carbon nanotube transparent conducting network films. *J. Phys. Chem. C*, 114, 9296–9300.
23. Moon, J. S., Park, J. H., Lee, T. Y., Kim, Y. W., Yoo, J. B., Park C. Y., Kim J. M., and Jim, K. W. (2005). Transparent conductive film based on carbon nanotubes and PEDOT composites. *Diamond Related Mater.*, 14, 1882–1887.
24. Liu, C., Zhang, Q., Stellacci, F., Marzari, N., Zheng, L., and Zhan, Zhan. (2011). Carbene-functionalized single-walled carbon nanotubes and their electrical properties, *Small*, 7, 1257–1263.
25. Andrew Ng, M. H., Hartadi, L. T., Tan, H., and Patrick Poa, C. H. (2008). Efficient coating of transparent and conductive carbon nanotube thin films on plastic substrates, *Nanotechnology*, 19, 205703–205707.

26. Xiao, G., Tao, Y., Lu, J., and Zhang, Z. (2010). Highly conductive and transparent carbon nanotube composite thin films deposited on polyethylene terephthalate solution dipping, *Thin Solid Films,* 518, 2822–2824.

27. Jonsson, S. K. M., Birgerson, J., Crispin, X., Greczynski, G., Osikowicz, W., Denier van der Gon, A. W., Salaneck, W. R., and Fahlman, M. (2003). The effects of solvents on the morphology and sheet resistance in poly (3,4-ethylenedioxythiophene)– polystyrenesulfonic acid (PEDOT–PSS) films, *Synth. Metals,* 139, 1–10.

Chapter 8

Effect of Edge Passivated by Hydrogen on the Transport Properties of Finite-Size Metallic Carbon Nanotube-Based Molecular Devices

Zongling Ding,[a,b,*] **Jun Jiang,**[c] **Huaizhong Xing,**[d] **Haibo Shu,**[b]
Yan Huang,[b] **Xiaoshuang Chen,**[b,*] **and Wei Lu**[b]
[a] *School of Physics and Material Science, Anhui University, Hefei 230039, China*
[b] *National Laboratory of Infrared Physics, Shanghai Institute for Technical Physics, Chinese Academy of Sciences, Yu Tian Road 500, Shanghai 200083, China*
[c] *Department of Theoretical Chemistry, School of Biotechnology, Royal Institute of Technology, Stockholm S-106 91, Sweden*
[d] *Department of Applied Physics, Donghua University, Ren Min Road 2999, Shanghai 201620, China*
*xschen@mail.sitp.ac.cn, zlding@ahu.edu.cn

In this chapter, the effect of edge passivated by hydrogen on the electronic and transport properties of the molecular devices of finite-size metallic carbon nanotubes (CNTs) is investigated by using density-functional theory in combination with Green's function method. Three types of hydrogenations are considered for the edge carbon atoms at the two open ends of the CNTs. The calculated

Advances in Nanodevices and Nanofabrication: Selected Publications from Symposium of Nanodevices and Nanofabrication in ICMAT2011
Edited by Qing Zhang, W. I. Milne, and Jianping Zou

ISBN 978-981-4364-54-6 (Hardcover), 978-981-4364-55-3 (eBook)
www.panstanford.com

energy gap between the highest occupied and the lowest unoccupied molecular orbitals decreases with increasing the length of the CNTs for the three hydrogen-passivated cases, respectively. Nonlinear current–voltage (*I–V*) curves and quantum conductance have been obtained in all junctions. It is shown that the electronic properties of the finite-size CNTs and the transport properties are sensitive to the passivation types of edge. With increasing the hydrogen passivation concentration of edge carbon atoms, it is indicated that the *I–V* characteristics have obviously the widening of bandgap and the decreasing of the quantum conductance.

8.1 Introduction

The discovery of CNTs has promoted numerous studies on its structures, properties, and potential applications [1,2] because of the remarkable structural, electrical, chemical and mechanical properties [3–5], and electronic transport as electronic devices [6–8]. It is well known that CNTs can display metallic, semiconducting and superconducting electronic transport. An infinitely long armchair nanotube is metallic, but energy gap is opened in the finite-length tubes [9]. When CNTs are shortened, their energy levels become quantized, which makes them be suitable for applications such as quantum dots and single-electron transistors [9, 10]. The reduced dimension can cause some important physical phenomena and the quantum effects become important. Because CNTs are formed by rolling up a graphene sheet into a tube, there are nonpassivated carbon atoms at the two open ends. The edge carbon atoms cannot be passivated [11–13] or can be passivated by hydrogen atoms [14–16]. The different passivation types of the edge atoms can have effect on the electronic properties of the CNTs. Furthermore, when the CNTs are sandwiched with the gold electrodes to form the molecular devices, the passivation types of the edge atoms have also influence on the connection configuration between the central CNTs and the metallic electrodes. The electronic and transport properties of ultra-short CNTs, and the device performance are strongly determined by the geometric structure of the tubes, the bias voltage and the contacts between CNTs and electrodes. Thus, in the present work we focus on the effect of the different passivation types of edge carbon atoms on the electronic structure and transport properties of ultra-short CNTs-based molecular devices. The length dependences of the bandgap and the conductance also have been studied.

8.2 Model and Method

The transport properties of junctions composed of finite, defect-free armchair single walled CNTs attached to Au(111) metallic electrodes have been calculated. The schematic diagrams of the ultra-short CNT based molecular devices with different edge passivation are shown in Fig. 8.1. Here, a metallic (3,3) tube is chosen as the prototype structure. Three passivation types for the open edges have been considered: (i) both ends are not passivated, (ii) both ends are passivated by hydrogen atoms, and (iii) one edge is not passivated, and another is passivated by hydrogen. A series of (3,3) ultra-short CNTs with *N*-carbon ($N = 4{\sim}22$) atomic layers along the tube axis with three types of edge passivations are considered. For simplicity, we label the CNTs as *N* O–O, *N* H–H and *N* O–H for the three types of model systems with different tube lengths, respectively. The notation *N* O–H indicates a finite segment of a (3,3) CNT, in which one of the two open edges is not passivated and another is saturated by H atoms with *N* carbon rings along its principal axis.

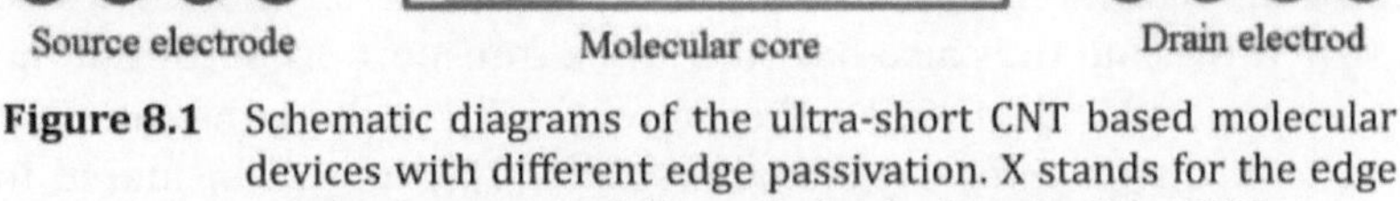

Figure 8.1 Schematic diagrams of the ultra-short CNT based molecular devices with different edge passivation. X stands for the edge passivation type and *N* stands for the length of the CNTs.

The geometry optimization and the electronic properties of the junctions are performed by using the density-functional theory in B3LYPlevel [17, 18] with mixed basis set (the 6-31G basis set for C and H atoms, and LanL2DZ basis set for Au atoms). The electronic transport properties are calculated by using the quantum chemistry for molecular electronics (QCME) code [19–21]. For our SWCNT based molecular junctions, the tunneling current density from the source to drain can be obtained with the following formula,

$$i_{1D} = \frac{2\pi e}{\hbar}\int_0^{\infty}[f(E-eV)-f(E)]T_{l'l}n^{S}(E)n^{D}(E)dE\ ,$$

where $f(E)$ is the Fermi distribution function, $n(E)$ is density of states and $T_{l'l}$ is the transmission probability describing the scattering process from the initial state $|l\rangle$ to the final state $|l'\rangle$, respectively.

8.3 Results and Discussions

It is well known that an infinitely long armchair nanotube is metallic, but in the finite-length tubes, the energy gap between the highest occupied molecular orbital (HOMO) and lowest unoccupied molecular orbital (LUMO) is opened due to the finite-size effect [9]. Several previous first principle calculations [10, 22, 23] predicted that the HOMO–LUMO gap increases as an oscillatory form with shortening the length [24–26]. The oscillation amplitude increases as the nanotube length decreases [27]. Our calculated energy gap between the HOMO and LUMO molecular orbitals for the considered CNT systems are showed in Fig. 8.2. One can see that the HOMO–LUMO gap is decreased as oscillation characteristics with the length of the CNT segment in the three types of edge passivation systems, respectively. With N increasing to ∞, the energy gap is expected to become zero. The result can be in part attributed to the quantum confinement along the tube length and curvature effects as the nanotube approaches to the zero-dimensional sizes. The confinement of the electronic wave functions caused by the finite-size limit of molecules has obviously influence on the electronic structures, and thus also has role in the transport properties of the nanomolecular devices. Furthermore, the CNTs show the different curvature with different diameters. The (3,3) tubes considered in our present work have a higher curvature. One can conclude that for the tubes with the lower curvature, the edge atoms should have sp^2 hybridization with a dangling σ bond. As the curvature increases, some electrons go to π orbital from σ orbital, and lead to sp^2–sp^3 hybridization. Furthermore, the results indicate that the region of the edge may also have sp^2–sp^3 re-hybridization because of the hydrogen passivation. The other reasons may come from the carrier distribution and the changes of the carrier concentration due to the hydrogen passivation types of edge atoms.

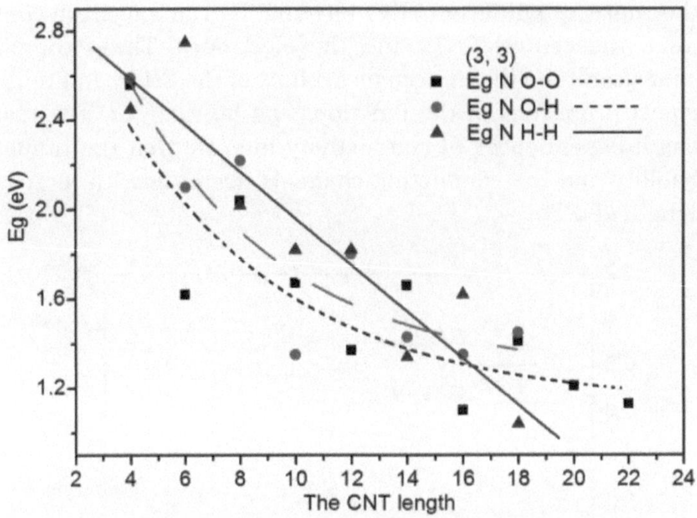

Figure 8.2 The HOMO–LUMO energy gaps as a function of the (3,3) SWCNT length in the *N* O–O, N O–H and *N* H–H systems are shown, respectively. The dots are the calculated results and the solid lines are the linear or exponential fittings of the data.

When the CNTs with different types of hydrogenation are sandwiched with the Au electrodes to form junctions, the hydrogenation of edge atoms can have effect on the electrode contact. As shown in Fig. 8.3(a–c), one can find the *I–V* and conductance-voltage (G/G_0–V, with G_0=$2e^2/h$ being the conductance quantum unit) curves for the 5 X–X (X stands the edge types, O for nonpassivated and H for hydrogen passivated) CNTs-based molecular junctions, respectively. Nonlinear *I–V* curves and quantum conductance have been obtained in all three junctions. From Fig. 8.3, it is clearly seen that *I–V* characteristics show a widening of bandgap after edge hydrogenation. By comparing the conductance of the 5 X–X systems, we can find that the conductance decreases with the increase of edge hydrogenation concentration. The result is explained by the enhancement of sp^3 hybridization, and the change of free-carrier concentration and distribution with hydrogen passivation at the edge carbon atoms of our tubes. It has been found both theoretically [28] and experimentally [29] that the quasi-free electrons tend to accumulate at the CNT's free termination. The edge hydrogenation of a CNT can lead to the modification of electronic structure, and the change of quasi-free electronic concentration and distribution.

Furthermore, the different edge termination can affect the contact between the central CNTs and the electrodes. These combined functions can tune the molecular orbitals of the CNTs, and thus the transport properties of the junctions can be changed. The results of length-dependences of conductivity indicate that the tunneling probability and the conducting channels decrease with increasing the length of CNTs.

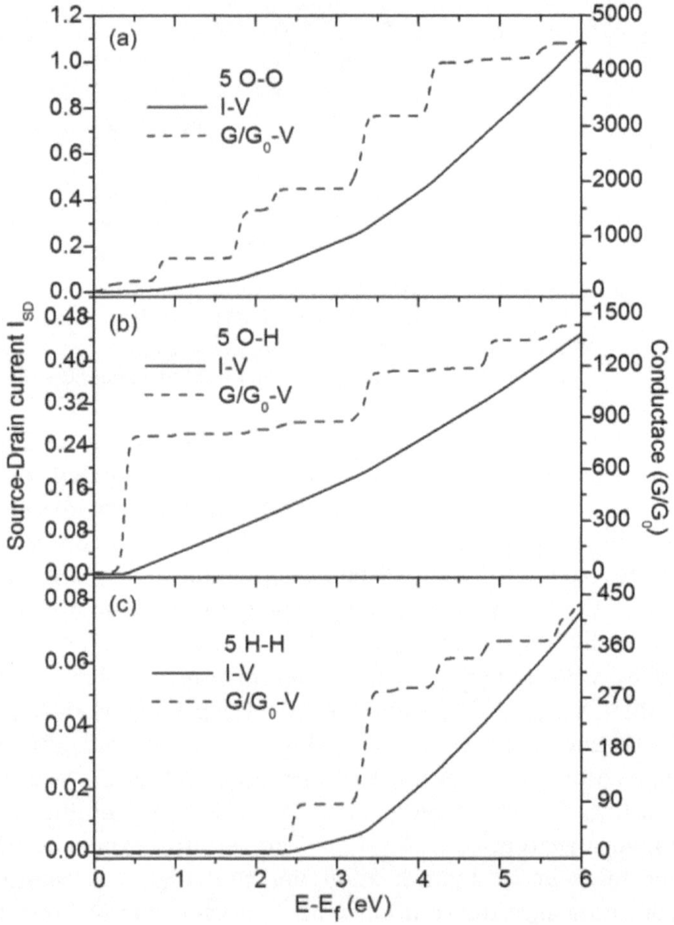

Figure 8.3 The current–voltage and quantum conductance (G/G_0, with G_0=$2e^2/h$ being the conductance quantum unit) curves of (a) 5 O–O, (b) 5 O–H and (c) 5 H–H CNT-based junctions (the solid and dashed lines correspond to the left and right axis, respectively).

8.4 Conclusion

To understand how the passivation types of edge atoms have effect on the electronic structure and transport properties in the finite-size metallic SWCNTs, we carry out extensively the DFT calculations in finite-length (3,3) CNTs with two open edges being clean or saturated by hydrogen. The calculated HOMO–LUMO energy gap increases with reducing the length of the CNT segment in the three edge passivation systems, respectively. Nonlinear *I–V* curves and quantum conductance have been obtained in all the junctions. The *I–V* characteristics show the widening of bandgap and the decrease of conductance obviously with the increase of edge hydrogenation concentration. The tunneling probability and the conducting channels decrease with increasing the length of CNTs. The feasibility of manipulating both electronic and transport properties of CNTs by controlling the edge passivation types may be useful in practical applications such as manipulating, functionalizing and etching of nanotubes.

Acknowledgment

This work was supported in part by the State Key Program for Basic Research of China (2007CB613206), the National Natural Science Foundation of China (10725418, 10734090, 10990104, 60976092, and 61006091), Key Fund of Shanghai Science and Technology Foundation (09DJ1400203, 10510704700, 10JC1416100), Science and Technology Commission of Shanghai Municipality (09dz2202200), the Creative Group (60221502), and the Knowledge Innovation Program of the Chinese Academy of Sciences. Computational resources from the Shanghai Supercomputer Center are acknowledged.

References

[1] Y. R. Yang, X. H. Yan, X. J. Shen, X. Y. Zhang, Y. Xiao. *J. Phys. Chem. C* 2010, 114: 7553–7557.

[2] J. Wang. *Electroanalysis* 2005, 17(1): 7–14.

[3] M. S. Dresselhaus, G. Dresselhaus, P. C. Eklund. *Science of Fullerenes and Carbon Nanotubes*, Elsevier: San Diego, 1996.

[4] H. Dai. *J. Acc. Chem. Res.* 2002, 35: 1035–1044.

[5] R. H. Baughman, A. A. Zakhidov, W. A. de Heer. *Science* 2002, 297: 787–792.

[6] R. Saito, M. S. Dresselhaus. *Physical Properties of Carbon Nanotubes*, Imperial College Press: London, UK, 1998.

[7] N. Sinha, J. Ma, J. T. Yeow, *J. Nanosci. Nanotechnol.* 2006, 6: 573–590.

[8] A. Naeemi, J. D. Meindl. *Annu. Rev. Mater. Res.* 2009, 39: 255–275.

[9] D. Y. Lu, Y. Li, S. V. Rotkin, U. Ravaioli, K. Schulten. *Nano Lett.* 2004, 4(12): 2383–2387.

[10] A. Rochefort, D. R. Salahub, P. Avouris. *J. Phys. Chem. B* 1999, 103: 641–646.

[11] K. Nakada, M. Fujita, G. Dresselhaus, M. Dresselhaus. *Phys. Rev. B* 1996, 54: 17954–17961.

[12] H. Lee, Y. W. Son, N. Park, S. Han, J. Yu. *Phys. Rev. B* 2005, 72: 174431(1)–174431(8).

[13] B. Huang, M. Liu, N. Su, J. Wu, W. Duan, B. Gu, F. Liu. *Phys. Rev. Lett.* 2009, 102: 166404(1)–166404(4).

[14] Y. Miyamoto, K. Nakada, M. Fujita. *Phys. Rev. B* 1999, 59: 9858–9861.

[15] Y. W. Son, M. L. Cohen, S. G. Louie. *Nature* 2006, 444: 347–349.

[16] L. Pisani, J. A. Chan, B. Montanari, N. M. Harrison. *Phys. Rev. B* 2007, 75: 064418(1)–064418(9).

[17] A. D. Becke, *J. Chem. Phys.* 1993, 98: 5648–5652.

[18] C. Lee, W. Yang, R. G. Parr, *Phys. Rev. B* 1988, 37: 785–789.

[19] J. Jiang, C. K. Wang, L. Yi. QCME-V1.1 (Quantum Chemistry for Molecular Electronics).

[20] C. K. Wang, Y. Luo. *J. Chem. Phys.* 2003, 119: 4923–4928.

[21] J. Jiang, M. Kula, Y. Luo. *J. Chem. Phys.* 2006, 124: 034708(1)–034708(10).

[22] J. Cioslowski, N. Rao, D. Moncrieff. *J. Am. Chem. Soc.* 2002, 124: 8485–8489.

[23] Y. Matsuo, K. Tahara, E. Nakamura. *Org. Lett.* 2003, 5: 3181–3184.

[24] H. Y. Zhu, D. J. Klein, T. G. Schmalz, A. Rubio, N. H. March. *J. Phys. Chem. Solids* 1998, 59: 417–423.

[25] A. Rochefort, D. R. Salahub, P. Avouris, *J. Phys. Chem. B* 1999, 103: 641–646.

[26] L. Liu, C. S. Jayanthi, H. Guo, S. Y. Wu, *Phys. Rev. B* 2001, 6403: 033414(1)–033414(4).

[27] X. M. Sun, S. Zaric, D. Daranciang, K. Welsher, Y. Lu, X. L. Li, H. J. Dai. *J. Am. Chem. Soc.* 2008, 130: 6551–6555.

[28] L. D. Filip, D. Nicolaescu, S. R. P. Silva. *J. Vac. Sci. Technol. B* 2006, 24(2): 874–880.

[29] P. Poncharal, Z. L. Wang, D. Ugarte, W. A. de Heer, *Science* 1999, 283: 1513–1516.

Chapter 9

A Study of Defect-Induced Electrical Breakdown on Individual Multi-Wall Carbon Nanotubes

Gopal Krishna Goswami and Karuna Kar Nanda*
Materials Research Centre, Indian Institute of Science, Bangalore, Karnataka 560012, India
*nanda@mrc.iisc.ernet.in

In this chapter, we report the electrical breakdown mechanism of long (40–50 μm) individual multi-wall carbon nanotubes (MWCNTs). Each MWCNT is patterned to have four metal electrodes by the electron beam lithography technique. The four electrodes have divided each MWCNT into three segments. The current–voltage measurements have been carried out for such devices to evaluate the resistance of each segment. It has been shown that the resistances of individual segments vary considerably. This can be due to the random distribution of defect density in different segments of MWCNTs. Hence, the highly defective segment would show the highest resistance. On applying a large bias across the ends of a carbon nanotube (CNT), we have observed that the highly resistive segment has been broken at a low breakdown power.

Advances in Nanodevices and Nanofabrication: Selected Publications from Symposium of Nanodevices and Nanofabrication in ICMAT2011
Edited by Qing Zhang, W. I. Milne, and Jianping Zou

ISBN 978-981-4364-54-6 (Hardcover), 978-981-4364-55-3 (eBook)
www.panstanford.com

9.1 Introduction

Since the discovery [1] in 1991, CNTs have drawn immense interest in research because of their excellent electrical properties. They have very high current carrying capacity (10^9 A/cm^2) [2] compared with conventional metallic wires, which do fail via electromigration [3]. CNTs having strong carbon–carbon covalent bonds prevent such atomic displacement under high current density. The very high current density has been experimentally observed for zero-bandgap metallic single-wall carbon nanotubes (SWCNTs), which show ballistic conduction [4] even at room temperature. However, these extraordinary electrical properties of CNTs are tied with their perfect defect-free structure. In general, presence of defects in the CNTs reduces [5] the electrical conductivity. Therefore, the nanoscale characterization of defects in CNTs is a worthy topic of discussion.

Typical defects such as kinks, holes and sliding between different walls have been observed through transmission electron microscope (TEM) [6]. The *in situ* wall-by-wall electrical breakdown [6] of MWCNTs depicts the fact that the breakdown might be initiated from the defect sites at high-bias condition where defect generation and migration is random along the CNT. Other characterization tools such as conducting atomic force microscopy (AFM) [7], scanning tunneling microscopy (STM) and spectroscopy [8] have been used to follow potential and current variations along a CNT to locate and characterize defects. Similarly, electrical characterization has also been employed to study the intrinsic as well as extrinsic defects in a CNT. Javey *et al.* showed that though short (electrode gap or equivalent effective tube length l is lesser than 100 nm) metallic CNTs without any significant phonon scattering exhibit ballistic conduction with the quantum conductance in the range of 2–4 e^2/h [9], the electrical breakdown occurs at lower biases as electrons encounter less back scattering and accelerate to high energies rapidly. The breakdown for CNTs of $l < 1$ μm has been related to Joule heating and resulting oxidation in air. However, for 3 μm-long CNTs, the electrical breakdown has been shown most likely at defect sites as the mean free path for elastic defect scattering is typically less than 3 μm. In general, the electrical breakdown of a carbon nanotube is predicted theoretically [10] at its middle due to poor heat dissipation through the contacts. However, experimentally, the electrical breakdown of bundle of CNTs has been observed [11] far away from the middle. This has been explained in terms of the presence of defects [11] at that site. Here, we have made an attempt to locate the defects along individual long CNTs to investigate the electrical breakdown through the measurement of resistances. We have electrically divided individual CNTs into four segments of equal l to compare the resistances of these segments. In general, the total resistance R_t of a device having a CNT–metal contact at its ends is a combination of the total CNT–electrode contact resistance R_c and the intrinsic resistance R_l of the CNT, which depends on l and the density of defects along the CNT. The resistance R_c for a particular metal–electrode contact may vary from one CNT to another but is constant for a given CNT. Therefore, the total resistance R_t is expected to be constant [12,13] (within the precision of the experiment) for different segments of identical l for a perfect defect-free CNT. Deviations from the above can give qualitatively the location of defects in a CNT. We have observed significant variation of resistances experimentally from different equi-segments of a MWCNT. When we have swiped large currents across the end electrodes of these devices, selective breakdown has been observed at the highly resistive segment. This has been correlated to the extent of defects present in the CNT that is responsible for electrical breakdown.

9.2 Experiment

Vertically aligned MWCNT bundles of more than 100 μm long have been synthesized by the pyrolysis of benzene and ferrocene in a quartz tube at a temperature of 900°C in a tube furnace. The synthesis procedure is similar as reported elsewhere [14]. The bundles of CNTs have grown at the side walls of the quartz tube. These are scratched carefully and separated out from the tube.

The bundles of CNTs have been dispersed in acetone to separate individuals and then drop casted on a Si/SiO_2 substrate. The individual CNTs are located in scanning electron microscopy (SEM) and assigned for electrode fabrication. The CNTs are significantly long (~40 μm) and ~100 nm in diameter. Conventional electron beam lithography technique has been employed to make metal contact on top of the CNTs. Pt (or Au) metal has been sputtered to make 100 nm-thick deposition to cover the CNT for better electrical transport. At least four electrodes have been made for individual CNT with identical l. Several devices have been fabricated with a variation of l from 2 to 6 μm. After the fabrication, devices are annealed at 180°C for 30 min for better CNT–electrode contact. The devices are fabricated for two- and four-probe resistance measurements in ambient at room temperature. Here, we have sourced currents and measure the voltage drop across the electrodes (*V–I* configuration). It is to be noted here that this *V–I* configuration would show voltage compliance with the current bias when electrical breakdown occurs.

9.3 Results and Discussions

We have carried out electrical measurements on several devices of similar configuration. The SEM image of one device has been shown in Fig. 9.1(a). A 40 μm-long CNT has been connected to four Pt electrodes separated by 6 μm (segment 1–2), 6 μm (segment 2–3) and 3.5 μm (segment 3–4) as measured from the SEM image. It is to be noted here that segment 3–4 has also been intended for 6 μm separation.

The two-probe resistance measurements for individual segments have been performed by applying current bias in the range of ±100 nA and the resistance has been calculated from the linear fit to the data points near the origin as shown in Fig. 9.1(b–d). The voltage–current variations in Figs. 9.1(b) and 9.1(d) are linear in nature and passing through the origin. This reveals the ohmic nature of the Pt–CNT contact. It is to be noted here that Pt (or Au) has higher work function than MWCNT, and hence these metals have been extensively used to make ohmic contacts [15] to CNT. Moreover, for our devices, we have found reasonable agreement between two- and four-probe resistance measurements. It is interesting to note that *VI* characteristics are linear for segments 1–2 and 3–4 but non-linear for segment 2–3. It is found that segment 2–3 has higher resistance (~314 MΩ) than segments 1–2 (~10.2 kΩ) and 3–4 (~7.23 kΩ). Hence, the non-linearity can be due to the Joule heating caused by very low current flow in a highly resistive segment. The resistances per unit length would be 1.70 kΩ/μm, 52.33 MΩ/μm and 2.06 kΩ/μm, respectively, for segments 1–2, 2–3 and 3–4. Such variation of resistances has also been observed for other devices. Another device has shown 1.9 kΩ/μm, 6.3 kΩ/μm and 27.5 kΩ/μm resistances for three segments of a CNT, separated by 2 μm distance. We believe that the variation of resistances at different segments of CNT can be due to the random distribution of defects along the CNT and does not depend on the contact resistance. A similar variation of resistance has also been reported [16] between any pairs of neighboring electrodes of a single MWCNT. However, detailed understanding is lacking.

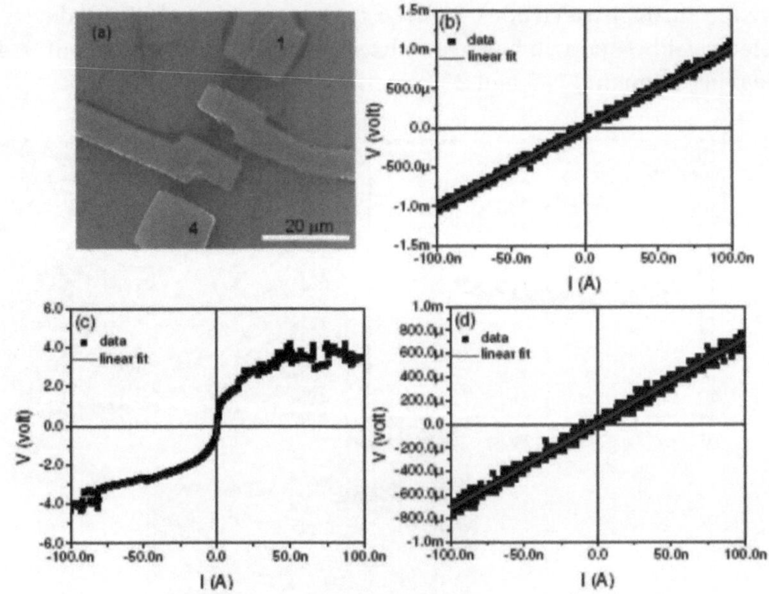

Figure 9.1 (a) SEM image of the fabricated device. The electrodes are numbered 1 to 4 from top to bottom. The two-probe V–I measurement of segment (b) 1–2, (c) 2–3 and (d) 3–4.

Therefore, the variations of resistances can be argued as intrinsic nature of the CNT as the contact resistance is expected to be invariable for different segments of CNT for a particular metal–CNT contact. However, when current is swiped from 0 to 2 mA across electrodes 1 and 4, we have observed electrical breakdown selectively at segment 2–3 well below 100 nA current as shown in Fig. 9.2(a). The resistance from Fig. 9.2(a) is found as 1300 MΩ, which is larger than the sum of the resistances of three segments. This can be realized by the fact that the total resistance is now the addition of the resistances of individual segments and the resistances of electrode covered region. Hence, highly resistive segment 2–3 has broken down even at low current. The breakdown is confirmed by the SEM micrograph as shown in Figs. 9.2(b) and 9.2(c). It is noteworthy to mention that segments 1–2 and 3–4 remain intact as seen in SEM as well as verified by low-bias *V–I* resistance measurements before and after the breakdown experiment. A similar result has also been observed for another device. The resistance of segment 3–4 (8.2 kΩ/μm) of this device is higher than that of segment 1–2 (5.4 kΩ/μm), and segment 2–3 is found to be electrically sorted. Upon high current swipe through electrodes 1 and 4 (similar to above) of the device, electrical breakdown has been observed selectively at segment 3–4 leaving segments 1–2 and 2–3 electrically intact.

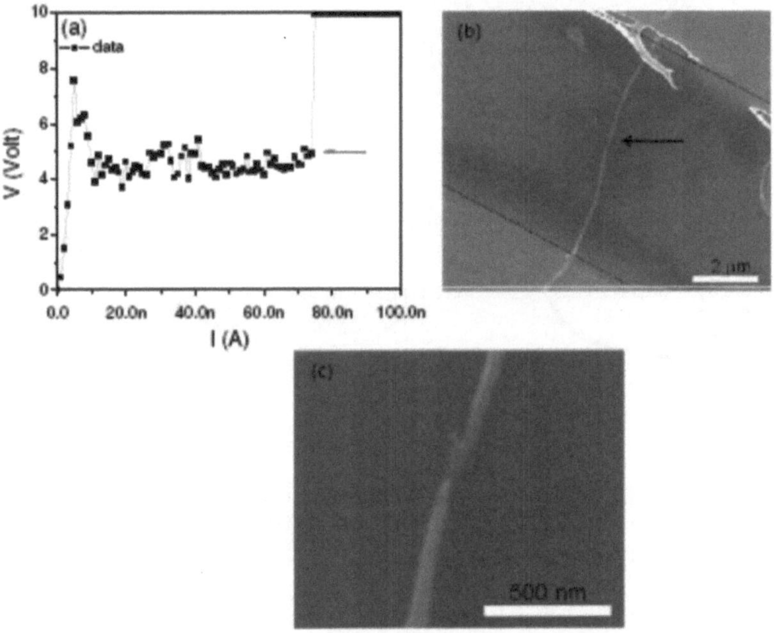

Figure 9.2 (a) The two-probe *V–I* measurement of segment 1–4 under high current bias. Breakdown has occurred below 80 nA as shown by an arrow head after which the curve shows its voltage compliance. (b) The SEM image of the device after breakdown. The electrodes are annotated for better clarity and the breakdown point has been shown by an arrow head. (c) The high-resolution SEM image of the broken CNT.

Hence, it can be concluded from the electrical breakdown that the highest resistance segment can have higher density of defects than the rest of the CNT under investigation. The breakdown at the defect site for a CNT longer than 3 μm can be supported by the fact that the length of the tube is larger than the mean free path for elastic defect scattering [9].

9.4 Conclusions

We have found experimentally that equispaced segments of a CNT are of varying resistances. On applying a high bias, the highest resistive segment has selective electrical breakdown. The highest resistive segment is expected to possess considerable local defects. Hence, our study offers insights into the possibility of electrical characterization of local defects along a CNT. This can be an alternative to highly complicated STM or AFM studies for electrical characterization of defects on individual CNTs for their applications in nanoelectronics.

References

1. Iijima, S. (1991). Helical microtubules of graphitic carbon, *Nature*, **354**, pp. 56–58.
2. Yao, Z., Kane, C. L., and Dekker, C. (2000). High-field electrical transport in single wall carbon nanotubes, *Phys. Rev. Lett.*, **84**, pp. 2941–2944.
3. Hauder, M., Gstottner, J., Hansch, W., and Schmitt-Landsiedel, D. (2001). Scaling properties and electromigration resistance of sputtered Ag metallization lines, *Appl. Phys. Lett.*, **78**, pp. 838–840.
4. Frank, S., Poncharal, P., Wang, Z. L., and Heer, W. A. (1998). Carbon nanotube quantum resistors, *Science*, **280**, pp. 1744–1746.
5. Bockrath, M., Liang, W., Bozovic, D., Hafner, J. H., Lieber, C. M., Tinkham, M., and Park, H. (2001). Resonant electron scattering by defects in single-walled carbon nanotubes, *Science*, **291**, pp. 283–285.
6. Huang, J. Y., Chen, S., Jo, S. H., Wang, Z., Han, D. X., Chen, G., Dresselhaus, M. S., and Ren, Z. F. (2005). Atomic-scale imaging of wall-by-wall breakdown and concurrent transport measurements in multiwall carbon nanotubes, *Phys. Rev. Lett.*, **94**, pp. 236802-1-4.
7. Dai, H., Wong, E. W., and Lieber, C. M. (1996). Probing electrical transport in nanomaterials: conductivity of individual carbon nanotubes, *Science*, **272**, pp. 523–526.
8. Ouyang, M., Huang, J. L., Cheung, C. L., and Lieber, C. M. (2001). Atomically resolved single-walled carbon nanotube intermolecular junctions, *Science*, **291**, pp. 97–100.

9. Javey, A., Guo, J., Paulsson, M., Wang, Q., Mann, D., Lundstrom, M., and Dai, H. (2004). High-field quasiballistic transport in short carbon nanotubes, *Phys. Rev. Lett.* **92**, pp. 106804-1-4.

10. Kuroda, M. A., Cangellaris, A., and Leburton, J-P. (2005). Nonlinear transport and heat dissipation in metallic carbon nanotubes, *Phys. Rev. Lett.*, **95**, pp. 266803-1-4.

11. Wei, Y., Jiang, K., Liu, L., Chen, Z., and Fan, S. (2007). Vacuum-breakdown-induced needle-shaped ends of multiwalled carbon nanotube yarns and their field emission applications, *Nano Lett.*, **7**, pp. 3792–3797.

12. Chiu, H. Y., Deshpande, V. V., Postma, H. W., Lau, C. N., Miko, C., Forro, L., and Bockrath, M. (2005). Ballistic phonon thermal transport in multiwalled carbon nanotubes, *Phys. Rev. Lett.*, **95**, pp. 226101-1-4.

13. Pablo, P. J., Navarro, C., Colchero, J., Serena, P. A., Herrero, J. Baro, A. M. (2002). Nonlinear resistance versus length in single-walled carbon nanotubes, *Phys. Rev. Lett.*, **88**, pp. 036804-1-4.

14. Mahanandia, P., and Nanda, K. K. (2008). A one-step technique to prepare aligned arrays of carbon nanotubes, *Nanotechnology*, **19**, pp. 155602 (7 pp).

15. Manohara, H. M., Wang, E. W., Schlecht, E., Hunt, B. D., and Siegel, P. H. (2005). Carbon nanotube Schottky diodes using Ti-Schottky and Pt-ohmic contacts for high frequency applications, *Nano Lett.* **5**, pp. 1469–1474.

16. Hsiou, Y. F., Yang, Y. J., Stobinski, L., Kuo, W., and Chen, C. D. (2004). Controlled placement and electrical contact properties of individual multiwalled carbon nanotubes on patterned silicon chips, *Appl. Phys. Lett.* **84**, pp. 984–986.

Chapter 10

Carbon Nanotube- and Graphene-Based Micro-Sensors and Reactors

Raluca Savu,[a,*] Rogério Valentim Gelamo,[a] Francisco Paulo Marques Rouxinol,[a] Alfredo Rodrigues Vaz,[a] Gustavo Monteiro da Silva Valente,[a] Alexander Flacker,[a,b] Mara Adriana Canesqui,[a] Angelo Luiz Gobbi,[c] and Stanislav Moshkalev[a]

[a] *Centro de Componentes Semicondutores—CCS, Universidade de Campinas—UNICAMP, C.P. 6101, Rua João Pandia Calógera, 90, Campinas, 13083-870, SP, Brazil*

[b] *Divisão de Microssistemas e Empacotamento, Centro de Tecnologia da Informação Renato Archer, Rodovia D. Pedro I Km 143.6, Campinas, 13069-901, SP, Brazil*

[c] *Laboratório de Microfabricação—Laboratório Nacional de Luz Sincrotron (LNLS), Rua Giuseppe Máximo Scolfaro 10.000, Pólo II de Alta Tecnologia, Campinas, SP, Brazil*

*raluca.savu@ccs.unicamp.br

10.1 Introduction

A large variety of materials have been synthesized recently in zero-dimensional (0D) or one-dimensional (1D) forms as nanoparticles and nanowires or nanotubes. These new materials (including metals, nitrides, sulfides, carbides and semiconducting oxides [1–3]) have attracted much attention as possible building blocks in electronic and photonic devices. The interest to these structures is due to high

Advances in Nanodevices and Nanofabrication: Selected Publications from Symposium of Nanodevices and Nanofabrication in ICMAT2011
Edited by Qing Zhang, W. I. Milne, and Jianping Zou

ISBN 978-981-4364-54-6 (Hardcover), 978-981-4364-55-3 (eBook)
www.panstanford.com

surface-to-volume ratios and quantum confinement effects, leading to unique optical and electrical properties and strong chemical reactivity. Nanoparticles are characterized by the highest surface-to-volume ratio and enhanced reactivity toward gases and liquids and thus have high potential for use in micro-sensors with reduced power consumption, enhanced sensitivity and fast time response.

The materials of choice for applications in gas sensing are wide-bandgap semiconductor metal oxides, which suffer strong changes in the conductance when oxidizing or reducing species are chemisorbed onto their surface [4,5]. Thin-film oxides are traditionally used in a large number of applications, but the trend toward miniaturization and the ever-growing need for devices with improved performance has induced an intense research activity in the area of nanostructure sensing elements [6].

However, the integration of low-dimensional materials in microelectronic devices, which requires controlled synthesis (predetermined position, size, etc.) or controlled deposition (basically, predetermined position and orientation), appears to be much more difficult for nanoparticles as compared with 1D materials like carbon nanotubes (CNTs). In contrast, micro-sensors in a chemiresistor configuration based on CNTs deposited precisely over metal electrodes from liquid solutions using the ac dielectrophoresis (DEP) process have been recently reported [7,8]. The number of the nanotubes deposited by this method can be readily controlled by the process conditions (deposition time, applied voltage, volume density of nanotubes), and individual nanotubes can be deposited as well. However, it should be noted that pristine nanotubes usually do not meet the basic gas sensor requirements of high and selective sensibility and short response and recovery times [9].

On the other hand, novel hybrid nanostructures, such as CNTs decorated with metal or metal oxide nanoparticles, proved to be very suitable for gas micro-sensing devices [7]. In this case, the functions important for gas sensing are separated between the two components as follows: (1) nanoparticles interact strongly with the reactive gas injected in the test chamber, and these reactions are known to generate electrical charges that can be then passed directly to CNTs. (2) Carbon nanotubes serve as (i) templates where nanoparticles can be grown or attached, (ii) local heaters that can reach temperatures as high as 400–500°C, due to Joule heating by the current passing through the nanotubes, (iii) low-resistance conductors (transducers) that allow reading the electrical signals produced in chemical interactions between particles and the gas. This separation of functions is important to provide a linear response of the sensor to detected gases, as the chemical interaction between gas and particles does not change the conductivity of nanotubes. This is in strong contrast with recently reported micro-sensors based on semiconductor nanowires that also are self-heated by Joule effect providing high reactivity toward reactive gases [7]. In the latter, the functions of chemical sensing, self-heating and signal transduction are not effectively separated, which may result in significant non-linearities in sensor response, as the resistance of a nanowire itself (and thus, self-heating) changes when its surface reacts with the gas. The decoration of nanotubes by both metal and semiconductor nanoparticles has been used in our previous works to increase sensitivity and to decrease response times, as well as their selectivity to different gases [7,9,10].

Micro-reactors that incorporate the sensors discussed in the previous paragraph have the advantage of combining nanostructure sensing materials with small working volume, offering better monitoring and feature control and increasing functionality of the devices [5]. They allow sensor miniaturization and open the possibility of use in many new applications, such as very compact, non-invasive pressure sensors, accelerometers and gas sensors [11,12]. Stimulated by developments in silicon micromachining, small multifunctional components have been successfully combined with traditional sensor technologies to enhance functionality in different sensing areas, for example, in micro flow sensors [5,11], micro temperature sensors [12], micro heaters [13] and micro pressure sensors [7].

10.2 Experimental Procedure

The reactors were designed in order to have a small volume while still being able to accommodate the substrate with the sensing heads and to fit to any vacuum/testing system (Fig. 10.1). The test chamber (inset in Fig. 10.1) is composed of a working cavity (100 μL volume), electric contact pins and an upper lid, all made in Kovar®. Connections for controlling pressure and gas composition in the working chamber are provided by 1/4″ Swagelok® tubes (1 mm internal diameter) placed at diametrically opposite positions. The body of the device was machined from a single metal piece, in order to avoid possible leaks in the system due to additional welding. The 0.5 mm Kovar® pins were electrically insulated from the metallic body of the reactor using commercial borosilicate sealing glass frit thermally treated at 800°C in a nitrogen atmosphere. After this process was completed, the chamber was gold plated and the reactor was hermetically sealed using a lid with Viton® o-ring.

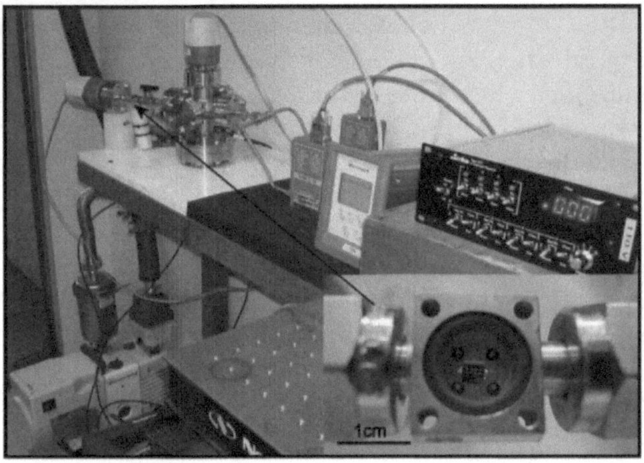

Figure 10.1 Image of the vacuum testing system that includes a reactor enclosing a complete sensor. The inset shows the open reactor with the sensing head electrically connected using the wire-bonding technique.

The sensors were fabricated by depositing multi-wall CNTs using DEP, onto oxidized silicon substrates with 1 μm-thick gold patterned electrodes. Gold layers 150 nm thick were sputtered over a 30 nm Ti adhesion layer and afterwards electrochemically thickened until 1 μm. Before the deposition, a gap of 0.5 μm wide and 5 μm deep was milled in the substrate by focused ion beam (NOVA 200 Nanolab — FEI Co.). This procedure allowed us to fabricate pressure sensors based on suspended nanostructures bridging the gap between the electrodes. The experimental parameters used for the DEF process were a frequency of 100 kHz, a 3 V peak-to-peak voltage and a deposition time varying from 15 to 75 s. The as-obtained sensors were mounted in the working chamber and the electrical connections were made using the wire-bonding technique (inset in Fig. 10.1). The sensors were characterized by field emission scanning electron microscopy (FE-SEM) and the electrical measurements were performed in pure nitrogen atmosphere using mass flow controllers of 100 sccm maximum flow (MB100 — MKS Instruments) and a 2410C Keithley source-meter in the working power interval from 0.001 to 0.02 W.

10.3 Results and Discussion

Figure 10.2 shows the schematic drawing for sensor configuration with pristine or decorated CNTs (Fig. 10.2a) and FE-SEM surface images of the gap milled between the electrodes (Fig. 10.2b) and sensors based on CNTs deposited by DEP over 1 μm-thick gold electrodes for 15 s (Fig. 10.2c) and 75 s (Fig. 10.2d). Note that individual tube was deposited under conditions of Fig 10.2c.

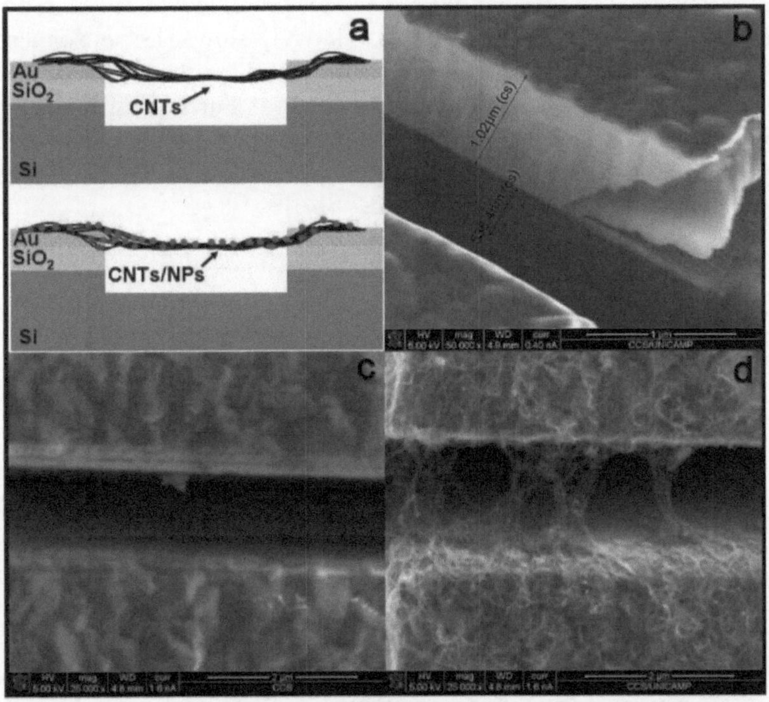

Figure 10.2 Schematic drawing of the suspended CNT-based gas sensors (a), FE-SEM images of the gap milled in the substrate by focused ion beam (b), and of the nanotubes deposited by DEP over the gold electrodes for 15 s (c) and 75 s (d).

As one can observe, the gap between the electrodes leaves the CNTs suspended (only in contact with the atmosphere), in this way increasing their sensibility. Previous studies have shown that when a bias is applied, much stronger heating of the nanotubes occur, due to Joule effect, for this configuration [26]. Then, the injection of the gas causes cooling of the nanotubes, and in turn, increase of the detected current, corresponding to the positive value of the temperature coefficient of resistance expected for metallic nanotubes [7,8]. Figure 10.3 shows the response of the sensor, obtained by depositing CNTs by 75 s (Fig. 10.2d), when the pressure was cycled between 5×10^{-3} and 4.5×10^{-1} mbar (this was the pressure reached when a N_2 flow of 100 sccm was introduced in the test chamber). The device was submitted to a number of vacuum/gas flow cycles and its response indicates good reproducibility during testing. The response time is fast (~2 s), compared with our previous results obtained with a 5 L test chamber (~10–15 s, not shown), but the recovery time is relatively slow. The sensor also exhibited good stability, both in vacuum and for the 100 sccm gas flow used, in the measurement time interval. Further testing using longer times, different working powers and other testing gases are needed in order to determine the optimum functioning conditions achievable with this kind of device.

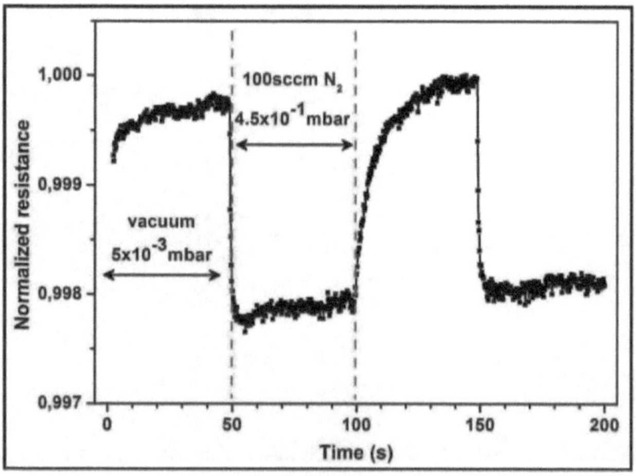

Figure 10.3 Sensor response (normalized resistance) as a function of time for the sensor obtained by depositing CNTs for 75 s over the gold patterned electrodes. The working power used was of 0.02 W.

In conclusion, technology of fabrication of novel micro-reactors containing CNTs in a chemiresistor configuration has been developed, and preliminary results obtained with suspended pristine CNTs show fast response to the injection of gases. Similar results were obtained also with multi-layer graphene flakes deposited using DEP [14]. Tests with decorated nanotubes and graphene are subject of future work.

References

1. C. N. R. Rao, F. L. Deepak, G. Gundiah, and A. Govindaraj (2003). *Prog. Solid State Chem.*, **31**, 5.
2. S. V. N. T. Kuchibhatla, A. S. Karakoti, D. Bera, and S. Seal (2007). *Prog. Mater Sci.*, **52**, 699.
3. A. Kolmakov and M. Moskovits (2004). *Annu. Rev. Mater. Res.*, **34**, 151.
4. N. Yamazoe, G. Sakai, and K. Shimanoe (2003). *Catal. Surv. Asia.*, **7**, 63.
5. W. C. Shin and R. S. Besser (2006). *J. Micromech. Microeng.*, **16**, 731.
6. E. Comini, C. Baratto, G. Faglia, M. Ferroni, A. Vomiero, and G. Sberveglieri (2009). *Prog. Mater. Sci.*, **54**, 1.
7. R. V. Gelamo, F. P. Rouxinol, C. Verissimo, A. R. Vaz, M. A. Bica de Moraes, and S. A. Moshkalev (2009), *Chem. Phys. Lett.*, **482**, 302.
8. R. V. Gelamo, F. P. Rouxinol, C. Verissimo, M. A. Bica de Moraes, and S. A. Moshkalev (2010). *Sens. Lett.*, **8**, 488.
9. A. Goldoni, L. Petaccia, S. Lizzit, and R. Larciprete (2010). *J. Phys. Condens. Matter.*, **22**, 013001.
10. A. Zevallos-Marques, M. J. Brasil, F. Iikawa, C. Verissimo, A. Abbaspourrad, S. A. Moshkalev, and O. L. Alves (2010). *J. Appl. Phys.*, **108**, 083501 (6p.).
11. P. Abgrall and N. T. Nguyen (2008). *Anal. Chem.*, **80**, 2326.
12. L. Dai, P. Soundarrajan, and T. Kim (2002). *Pure Appl. Chem.*, **74**, 1753.
13. R. M. Tiggelaar, P. van Male, J. W. Berenschot, J. G. E. Gardeniers, R. E. Oosterbroek, M. H .J. M. de Croon, J. C. Schouten, A. van den Berg, and M. C. Elwenspoek (2005). *Sens. Actuators A*, **119**, 196.
14. F. P. Rouxinol, R. V. Gelamo, R. G. Amici, A. Vaz, and S. A. Moshkalev (2010). *Appl. Phys. Lett.*, **97**, 253104 (3p.).

Chapter 11

Carrier Statistics Modeling of p-Type Graphene Nanoribbons

Seyed Mahdi Mousavi,[a] Mohammad Taghi Ahmadi,[a,*] Jeffrey Frank Webb,[b] Zaharah Johari,[a] Azizah Amin,[a] and Razali Ismail[a]

[a] *Faculty of Electrical Engineering, Universiti Teknologi Malaysia, Skudai, Johor, Malaysia*

[b] *School of Engineering and Science, Swinburne University of Technology Sarawak Campus Kuching, Sarawak, Malaysia*

*ahmadiph@gmail.com

Graphene nanoribbons (GNRs) are quasi one-dimensional nanostructures. This chapter discusses carrier statistics for p-type GNRs. This is one of the fundamental properties of these structures and is of particularly interest with respect to applications of GNRs in electronic devices. Carrier concentration expressions for GNR in both the degenerate and non-degenerate regimes will be presented based on a mathematical model that captures the essential behavior. The model takes into account the fact that in the degenerate regime the Fermi–Dirac integral varies with carrier concentration, whereas in the non-degenerate regime it is independent of carrier concentration but depends strongly on temperature.

Advances in Nanodevices and Nanofabrication: Selected Publications from Symposium of Nanodevices and Nanofabrication in ICMAT2011

Edited by Qing Zhang, W. I. Milne, and Jianping Zou

ISBN 978-981-4364-54-6 (Hardcover), 978-981-4364-55-3 (eBook)

www.panstanford.com

11.1 Introduction

As we approach the end of semiconductor road map for conventional devices and quantum effects start to become increasingly important, it is expected that carbon-based devices such as carbon nanotubes (CNTs), nanowires, and GNRs will help resolve many downscaling problems (such as unwanted quantum tunneling and leakage currents). This may lead to a new paradigm for electronic device design and be of particular importance to the creation of nanoscale transistors [1–2]. Although significant progress has been achieved in the fabrication of GNR devices, [3–4] many details are still need to be worked out in GNR analytical modeling, as the device physics is still relatively unexplored. Naturally GNR is a two-dimensional (2-D) system by reducing the width the GNR carriers are confined in quasi-1D system similar to a CNT. The electrical properties of graphene can be transformed from 2-D sheets to quasi-1-D nanoribbons [5]. A GNR can be thought of as an unzipped single-wall CNT (SWCNT) as shown in Figures 11.1 and 11.2.

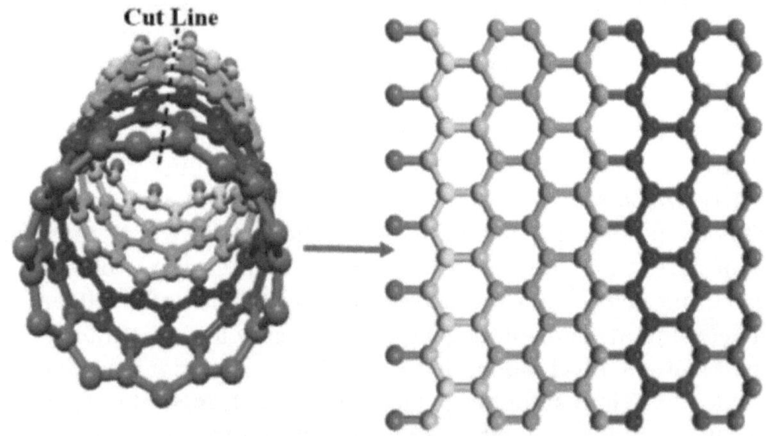

Figure 11.1 An armchair (m,m) SWCNT cut can spread out into either an (m,m) or an $(m - 1/2, m - 1/2)$ armchair GNR.

There are two different ways to unfold a SWCNT that result in two different classes of GNRs; each way has a different topology referred to as armchair or zigzag, names which refer to the shape of the edges (the horizontal edges in Figures 11.1 and 11.2) in each case. An armchair GNR can be formed by unzipping the SWCNT along the axial direction. Any given m, m armchair edge section from the cut along the SWCNT can spread out into either an (m, m) or an $(m - 1/2, m - 1/2)$armchair GNR—see Figure 11.1. A zigzag GNR is formed by cutting along a different line, as shown in Figure 11.2. Now, any given $(m, 0)$ zigzag from the cut line can spread out into either an (m – 1, 0) or $(m - 1/2, 0)$ [6].

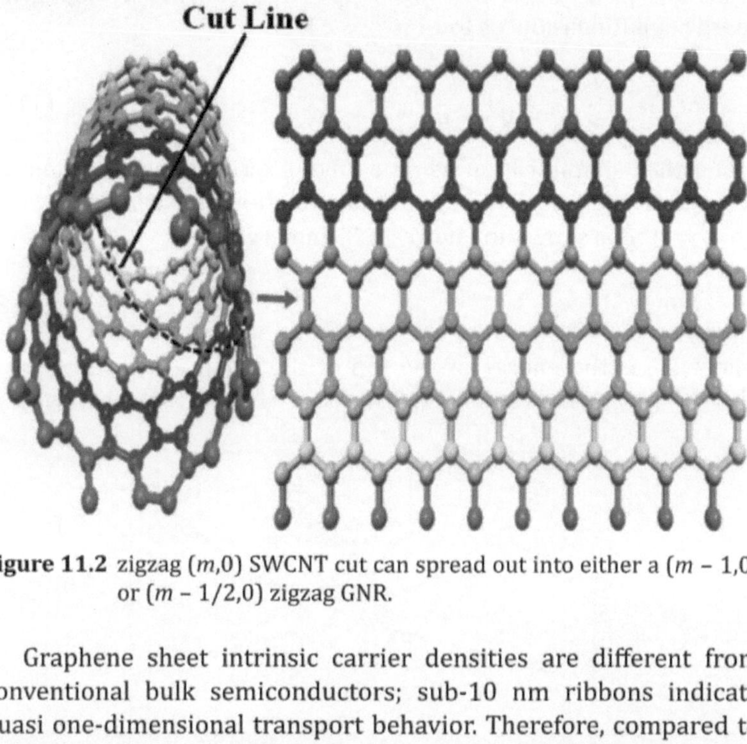

Figure 11.2 zigzag $(m,0)$ SWCNT cut can spread out into either a $(m - 1,0)$ or $(m - 1/2,0)$ zigzag GNR.

Graphene sheet intrinsic carrier densities are different from conventional bulk semiconductors; sub-10 nm ribbons indicate quasi one-dimensional transport behavior. Therefore, compared to conventional bulk semiconductors, more basic quantum mechanical principles need to be invoked in order to study the physics of the carriers [7,8].

In this chapter, we focus on a theoretical study of GNRs with band structure taken as a fundamental basis for calculating the carrier statistics of p-type GNRs.

11.2 Graphene Nanoribbon Band Structure

The band energy throughout the entire Brillouin zone of the valence band for graphene is

$$E(k) = -\frac{3t}{2} a_{c-c} \sqrt{\beta^2 + k_y^2} \tag{11.1}$$

where a_{c-c} is the distance between adjacent carbon atoms, which is approximately 1.42 Å, t = 2.7 (eV) is the nearest neighbor C-C tight binding overlap energy, k_y is the wave vector component along the y direction and β_x and β_y show the confinement in the x and y directions [9–10], and $\beta^2 = \beta_x^{\,2} + \beta_y^{\,2}$. After some rearranging the energy equation reduces to

$$E(k) = -\frac{3t}{2} a_{c-c} \beta \sqrt{1 + \frac{k_y^{\,2}}{\beta^2}} \tag{11.2}$$

Further simplification using a Taylor series approximation for the square root results in a parabolic relation between energy and wave vector as shown in Figure 11.3, and is given by

$$E \approx E_{v_0} - \frac{\hbar^2 k_y^2}{2m^*} \tag{11.3}$$

where E_{v_0} is the energy of the top of the valence band and $m^* = (2/9)\hbar^2/(ta_{c-c})^2 E_g$ is an effective mass in which E_g is the bandgap.

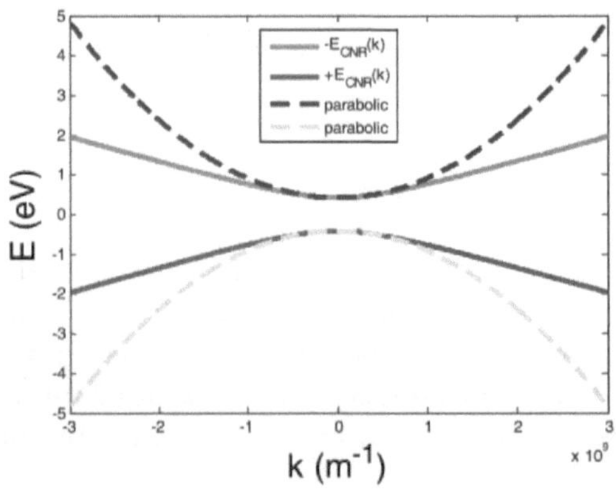

Figure 11.3 The band structure of a GNR near the minimum energy is parabolic.

In the parabolic part of the band energy, which is where we focus our attention, the behavior of the GNR (carrier concentration, velocity of carrier and carrier transport) can be explained by Fermi–Dirac integral formalism. The density of states can is given by [10, 11]

$$D(E) = \left(\frac{-m^*}{2\pi\hbar^2}\right)\left(\frac{2m^*(E_{vo} - E)}{\hbar^2}\right)^{-1/2} \tag{11.4}$$

By increasing the width of the GNR the bandgap decreases as shown in Figure 11.4. For sub-10 nm widths, a GNR behaves like a semiconductor. In contrast, widths above 10 nm exhibit metallic GNR. This is in line with an empirical equation that indicates that the bandgap in a semiconducting GNR is a function its width W. The equation relating the width to the bandgap energy, using the fact that the wave vector is quantized in across the width of the ribbon, is [1, 2]

$$E_g = \frac{0.8 \text{ eV} \cdot \text{nm}}{W(\text{in nm})}$$

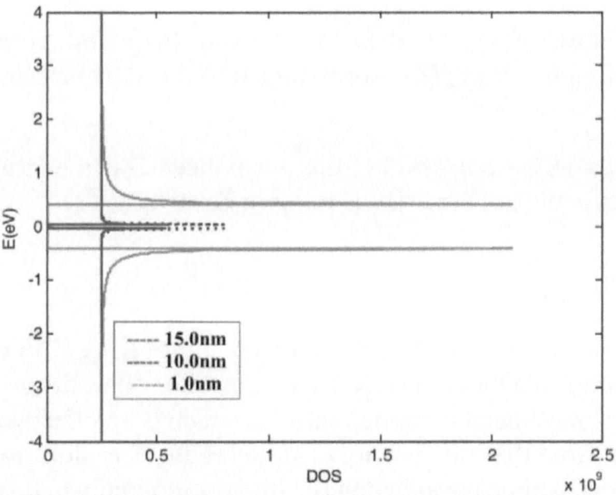

Figure 11.4 Density of states for a GNR with variation of width.

With this modification m^*, longitudinal effective mass, depends on the width of the ribbon and can be expressed as

$$m^* = \frac{2}{9}\frac{\hbar^2}{(ta_{c-c})^2}\left[\frac{0.8 \text{ eV} \cdot \text{nm}}{W(\text{in nm})}\right] = \frac{0.092m_o}{W(\text{in nm})} \tag{11.5}$$

The approximation leads to Fermi–Dirac integrals in the parabolic band structure for carrier concentration, velocity, and current, which are of the same form as used for the conventional channel of one-dimensional devices.

11.3 Equilibrium Carrier Statistics

The number of carriers per volume with energies between E and $E + dE$ at temperature T is calculated from $D(E)\,f\,(E)dE$ for electrons, and for holes, of interest here, by $D(E)\,[1 - f\,(E)]dE$, where $f(E)$ is the Fermi–Dirac distribution function given by

$$F(E) = \frac{1}{e^{(E-E_F)/(k_BT)} + 1} \tag{11.6}$$

where E_F is the Fermi energy and k_B is Boltzmann's constant.

Therefore the total carrier concentration p in a band is obtained simply by integrating over the valence band, so that

$$p = \int_{E_{bottom}}^{E_{v0}} D(E)[1 - f(E)]dE \tag{11.7}$$

By substituting the density of state $D(E)$ and Fermi–Dirac distribution function $f(E)$ expressions into the above we have

$$p = N_v \Im(\eta) \tag{11.8}$$

where $\eta = (E_{v0} - E_F)/(k_BT)$ is the normalized Fermi energy, $\Im$ is a special case of the Fermi–Dirac integral given by

$$\Im(\eta) = \int_0^{\infty} \frac{1}{e^{(x-\eta)} + 1} dx \tag{11.9}$$

where $x = (E - E_{v0})/k_BT$, and the terms multiplying $\Im$ in equation (11.8) after algebraic manipulation of the relevant proceeding equations have been absorbed into N_v, which is an effective density of states. Note that the density of states reduces rapidly away from the valence band edge allowing the integration limits to be replaced by those in equation (11.9) with little loss of accuracy.

Under non-degenerate conditions, the one in the denominator of the integrand in equation (11.9) can be neglected and the Fermi integral becomes an exponential (like a Maxwellian distribution) given by

$$\Im(\eta) = e^{\eta} \tag{11.10}$$

In the strongly degenerate regime, the Fermi integral transforms to

$$\Im(\eta) = \eta \tag{11.11}$$

In the non-degenerate regime the Maxwellian approximation carrier concentration only depends on the temperature. Therefore for hole concentration we obtain

$$p = N_v e^{-\left(\frac{E_{v0}-E_F}{K_B T}\right)} \tag{11.12}$$

This simplified distribution function is extensively used in determining the transport parameters. This simplification is true for non-degenerately doped semiconductors. However, most nanoelectronic devices these days are degenerately doped. Hence any design based on the Maxwellian distribution is not strictly correct and often leads to errors in the interpretation of experimental results. To avoid these problems, however, for the degenerate case we have

$$p = \left[\frac{8m^*(E_F - E_{v0})}{\pi^2 \hbar^2}\right]^{\frac{1}{2}} \tag{11.13}$$

Figure 11.5 shows that the Fermi integral is closely approximated by e^η when $\eta \leq -3$, the non-degenerate regime, and by η for $\eta \geq 6$, the degenerate regime.

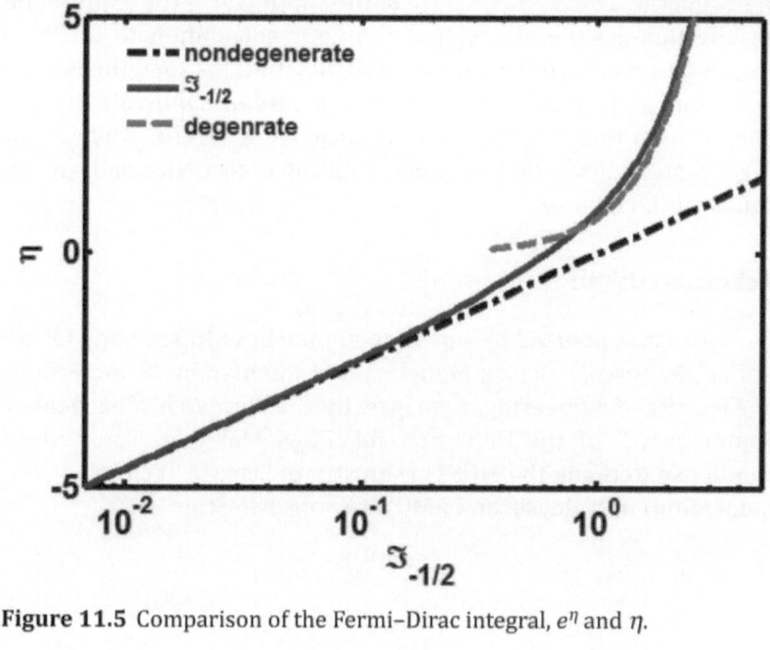

Figure 11.5 Comparison of the Fermi–Dirac integral, e^η and η.

As expected in the non-degenerate regime $(E_v - E_F)$ as a function of p/N_{v0}, using and , is given by

$$(E_{v0} - E_F) = k_B T \ln (p/N_{v0}) \tag{11.14}$$

Here, $(E_{v0} - E_F)$ is a weak (logarithmic) function of carrier concentration but varies linearly with temperature. However, for strongly degenerate statistics, the Fermi energy is independent of temperature and is a strong function of carrier concentration, since, from equations (11.6) and (11.10),

$$(E_v - E_F) = [N_v p/2]^2 \tag{11.15}$$

The Fermi energy is proportional to p^2.

11.4 Conclusion

Our analysis has shown that for quasi one-dimensional semiconductor-like devices, such as a nanowire transistors and CNTs, the Fermi–Dirac integral is given by e^{η} in the non-degenerate approximation, and by η in the degenerate approximation.

Induced and doped carrier densities in most nanoscale devices are now in the degenerate regime, thus generating a great interest in degenerate statistics. Because of the simplicity of the expressions for the non-degenerate regime, it is not uncommon to interpret results based on non-degenerate statistics, but this sometimes leads to erroneous results, particularly in the case of nanoscale devices. Therefore in this chapter, we have included a careful study of the degenerate regime that is more applicable to GNRs and similar nanoscale structures.

Acknowledgment

The work is supported by a post-doctoral fellowship scheme for the project Nanoscale Device Modeling and Simulation, at the Faculty of Electrical Engineering, managed by the Research Management Center (RMC) of the Universiti Teknologi Malaysia. The authors would like to thank the MOSTI (Ministry of Science Technology and Innovation) in Malaysia and RMC for sponsoring this work.

References

[1] K. Natori, Ballistic metal–oxide–semiconductor field effect transistor, *J. Appl. Phys.*, 76, 4879–4890, 1994 (October).

[2] F. Assad, Z. Ren, D. Vasileska, S. Datta, M. Lundstrom, On the performance limits for Si MOSFET's: A theoretical study, *IEEE Trans. Electron Devices*, 47, 232–240, 2000 (January).

[3] X. Li, X. Wang, L. Zhang, S. Lee, H. Dai, Chemically derived, ultrasmooth, graphene nanoribbon semiconductors, *Science*, 319, 2008.

[4] X. Wang, Y. Ouyang, X. Li, H. Wang, J. Guo, H. Dai, "Room-temperature all-semiconducting sub-10-nm graphene nanoribbon field-effect transistors, *Phys. Rev. Lett.*, 100, 206803, 2008.

[5] A. K. Geim, K. S. Novoselov, The rise of graphene, *Nat. Mater.*, 6, 183–191, 2007.

[6] Q. Yan, B. Huang, J. Yu, F. Zheng, J. Zang, J. Wu, B.-L. Gu, F. Liu, W. Duan, Intrinsic current–voltage characteristics of graphene nanoribbon transistors and effect of edge doping, *Nano Lett.*, 7, 1469–1473, 2007.

[7] T. Fang, *et al.*, Carrier statistics and quantum capacitance of graphene sheets and ribbons, *Appl. Phys. Lett.*, 91, 092109, 2007.

[8] V. Sverdov, *et al.*, Current transport models for nanoscale semiconductor devices, Mater. *Sci. Eng. Reports*, 58(6), 228–270, 2008.

[9] J. W. G. Wildöer, L. C. Venema, A. G. Rinzler, R. E. Smalley, C. Dekker, Electronic structure of atomically resolved carbon nanotubes, *Nature* (London), 391(6662), 59–62, 1998.

[10] R. A. Jishi, D. Inomata, K. Nakao, M. S. Dresselhaus, G. Dresselhaus, Electronic and lattice properties of carbon nanotubes, *J. Phys. Soc. Jap.*, 63(6), 2252–2260, 1994.

[11] M. T. Ahmadi, *et al.*, The ultimate ballistic drift velocity in carbon nanotubes, *J Nanomater*, 2008, 769250, 2008.

[12] M. T. Ahmadi, *et al.*, Analytical study of drift velocity in P-type silicon nanowires, *Nanosci. Nanotechnol.*, 1136, 264–268, 2009.

Chapter 12

Various Tunable Photonic Crystal/ Quasi-Photonic Crystal Structures Fabricated by Reconfigurable Interference System

Haitao Dai,[a] Xiaowei Sun,[b,*] Yanjun Liu,[b] and Dan Luo[b]
[a] *Department of Applied Physics, School of Science, Tianjin University, Tianjin 300072, China*
[b] *School of Electrical and Electronic Engineering, Nanyang Technological University, Nanyang Avenue, 639798 Singapore*
*exwsun@ntu.edu.sg

In this chapter, we demonstrate a reconfigurable interference lithography system (RILS) based on digital mirror device (DMD), which enables the fabrication of photonic crystals (PCs) or quasi-PCs structures with various structural lattices constant readily. The interference system is composed of a DMD chip, a concave lens and a convex lens. The interferential patterns are recorded in the polymer-dispersed liquid crystal (PDLC) materials, which can be electrically tuned and switched. We also propose another reconfigurable system combined with an axicon lens to achieve large effective area.

Advances in Nanodevices and Nanofabrication: Selected Publications from Symposium of Nanodevices and Nanofabrication in ICMAT2011
Edited by Qing Zhang, W. I. Milne, and Jianping Zou

ISBN 978-981-4364-54-6 (Hardcover), 978-981-4364-55-3 (eBook)
www.panstanford.com

12.1 Introduction

Photonics crystals have emerged as very important structures in a variety of optical fields, such as optical communication [1] and integral photonic circuit [2,3]. Most recently, structures of PCs have also attracted a lot of efforts because of their applications in extracting light from light emitting diode (LED) [4] or organic LED (OLED) and enhancing light absorption in solar cells [5]. The fabrication of 2D or 3D PC structures has been the greatest challenge in the past decades. Many procedures have been developed to fabricate photonics structures, such as deep ultraviolet photolithography, a standard tool in electronics industry; e-beam lithography [6]; multi-photon direct laser writing [1]; and holographic photolithography (multi-beam interference lithography) [7,8]. Generally, multi-beam interference lithography is used comprehensively owing to its simple configurations, rapid fabrication, low cost and throughput scale. The 2D or 3D interferential intensity profiles formed by multi-beam are transcribed into photoresist by common photolithographic procedure. According to this method, 2D/3D photonic template can be achieved with nearly defect-free structures, micron or sub-micron scale lattice over large substrate.

Most of the reported configurations to fabricate 2D/3D PCs by multi-beam interference are meant to achieve beamlets by pre-design prism [9,10], diffraction optical elements (DOEs) [11–13] or splitting beams directly by bulk optical elements [14]. It is not easy to vary the lattice constant of PCs once the prism or mask or bulk optical configuration has been defined. To address this issue, some researchers also fabricated the 2D or 3D PCs by imaging simulated phase or intensity patterns by multi-beam interference displayed on the spatial light modulator in the record plane [15,16]. Nevertheless, the lattice constant of fabricated sample was limited by the pixel width of the device. Here we present a reconfigurable multi-beam generated system according to a DMD-based reconfigurable interference system. Based on this reconfigurable system, the number of beamlets and the individual incident angle of incident beam can be tuned readily. The generated patterns are recorded in the PDLC materials, which can be electrically switched [17–19]. We demonstrated three beams interference with different incident polar angles and azimuth angles, respectively. We also fabricated quasi-crystals structures in PDLC by five beams' interference.

12.2 Methodology

The reconfigurable multi-beam lithographic system is based on the DMD chip, which was developed as an optical engine by Texas Instrument to achieve high-quality projection display with fast response time. It is also used comprehensively as an amplitude spatial light modulator to generate kinds of pre-designed intensity patterns. The DMD's amplitude modulated property therefore enables effective multibeams generation with various parameters such as numbers, radius and shape, etc. In addition to DMD chip, the interference system is composed of a concave lens and a convex lens as well.

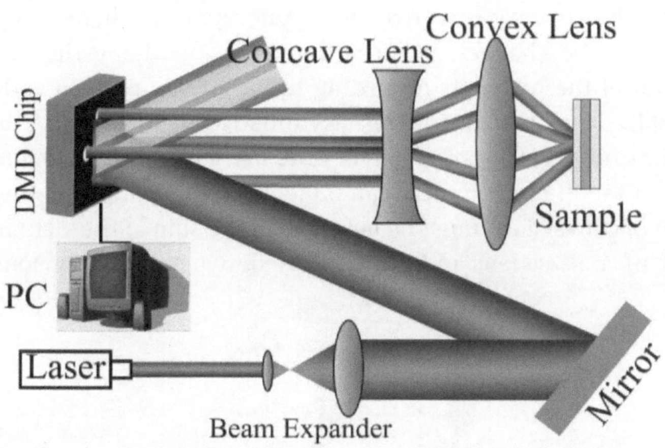

Figure 12.1 Schematic of RILS.

Figure 12.1 shows the schematic of the DMD-RILS, the light beam emitted from laser was impinged on the DMD chip after expanded and collimated. The DMD chip was controlled via VGA port of a personal computer; hereafter, the incident beam will be split into several beamlets by the driven pattern employed on DMD chip with additional energy loss due to the amplitude modulated mechanism of DMD. Before interference at the sample to form PCs, the formed beamlets pass through a concave lens and a convex lens. The concave lens is used to extend the spacing between the beamlets to achieve various incident angles to tune PCs pitch. As shown in Figure 12.2, the final convergence angle θ of the beamlets with fixed incident position h_1 will be determined by the combination of concave lens and convex lens. With simple geometric optics calculation, we can calculate the convergence angle of the incident beam parallel with respect to optical axes in terms of the geometric parameters of lenses as follows:

$$\tan\theta = \frac{h_2}{v} = \frac{h_1(f_1 - f_2 + d)}{f_1 f_2} \qquad (12.1)$$

Here f_1 and f_2 are the focal lengths of the concave lens and the convex lens, respectively; u and v represent the object distance and the image distance of the convex lens, respectively. As we can see from Eq. 12.1, as for a given incident beam, the convergence angle will vary with different distance d between concave and convex lens under fixed focal lengths situation. Thus it can be tuned readily by moving the concave or convex lens. One the other hand, the angle θ also can be changed by programmably modifying the incident position of the beamlets according to the driven pattern coded on the DMD chip, which enables asymmetric beamlets distribution and thus makes it possible to fabricate more complex structures. To ensure that all the beamlets can be focused to focal plane, the focal length of convex lens must be larger than the sum of the virtual focal length of concave lens and the distance between these two lenses.

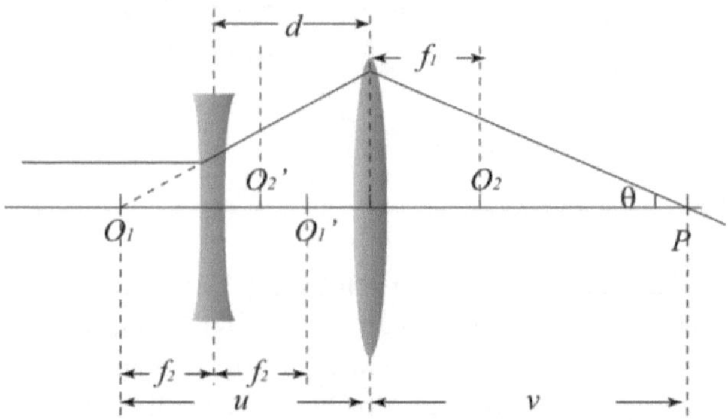

Figure 12.2 The configuration of convex and concave lens to tune convergence angle of the incident beamlets.

12.3 Theoretical Description

Many efforts have been engaged in the simulation of multi-beam interference. Here we briefly explore the multi-beam lithographic technique. In general, the intensity distribution $I(r)$ of the multi-beam interference results from the superposition of plane coherent waves and can be expressed as

$$I(r)=\sum_{n,m}\vec{E}_n\exp\left[-i(\vec{k}_n\cdot\vec{r}+\delta_n)\right]\cdot\vec{E}_m^*\exp\left[i(\vec{k}_m\cdot\vec{r}+\delta_m)\right] \quad (12.2)$$

where $\vec{E}$ is the optical electric field vector (the star symbol denotes complex conjugate), $\vec{r}$ is the radius vector, $\vec{k}$ is the wave vector, and $n = m$ represents the number of interfering beams. The phases of beams are given by δn, $m = 0$ in our situation. Figure 12.3 shows the driven patterns employed on the DMD chip and the correspondent intensity profiles.

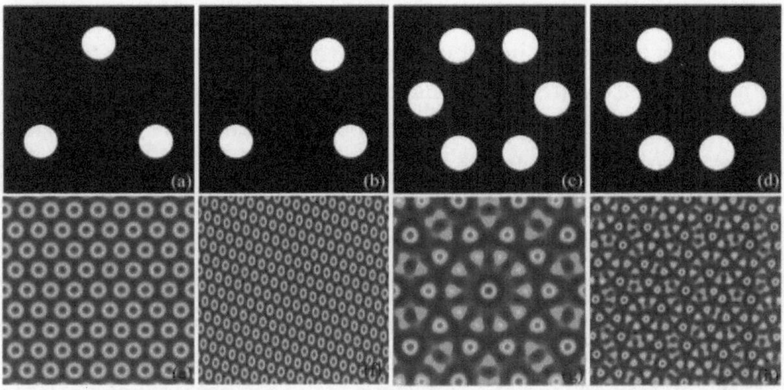

Figure 12.3 The driven patterns with three beams (a, b) and five beams (c, d) and the correspondent interference profiles(e–h); (b) and (d) are the three- and five-beam-driven patterns with positions slightly varied.

12.4 Experimental Results

To achieve the multi-beam according to DMD chip, it is readily to form the driven pattern and drive the DMD device via VGA port according to the personal computer. Figures 12.3(a–d) shows the driven pattern designed in computer to generate three or five beams with different geometric parameters. In our experiment, a collimated laser beam (532 nm) after expended to ensure that the radius of beam is enough to cover the surface of DMD impinges onto the surface of a DMD chip (1024 × 768 pixels with a pitch of 10.8 μm). The sample is placed at the re-superposition plane of the RILS to record the interferential profiles. The syrup used is the same as described in Ref. [13]: trimethylolpropane triacrylate (TMPTA); cross-linking monomer, *N*-vinylpyrrollidone (NVP); surfactant, octanoic acid (OA); coinitiator, *N*-phenylglycine (NPG); and photoinitiator, rose bengal (RB); all from Aldrich. In prepolymer, the ratio of TMPTA/NVP/OA/NPG/RB was 62/25/10/2/1 by weight. The LC used was E7 (Merck) with an ordinary refractive index of no = 1.521, and a birefringence of Δn = 0.225. The prepolymer and LC were mechanically blended in dark conditions according to the appropriate weight ratio at 65°C (higher than the clearing point of the LC E7) to form a homogeneous mixture. In this prepolymer/LC mixture, the LC concentration was about 35 wt%. A droplet of the mixture was sandwiched between two pieces of indium tin oxide (ITO) coated glass. The cell gap was about 30 μm. The curing intensity of individual beams before the sample was 12 mW/cm^2 and the exposure time was 120 s. After exposure, the samples were further cured for 5 min using a UV lamp to ensure the complete polymerization of the prepolymer.

Figure 12.4 shows the structures fabricated by the proposed RILS with three beamlets. In our experimental, the values of focal lengths of the concave and the convex lens are both equal to 10 cm with about 40 cm distance between them, the interval of beamlet's center with the axis of lens (optical axis) is 3 mm, which correspond to a about 7° convergence angle resulting in about 2 micron structure lattice constant. Figure 12.4(a) depicts the scanning pictures of the sample by removing LC from polymer matrix with AFM, and the corresponding far-field diffraction pattern is shown in Figure 12.4(b). Six points refer to the first diffraction order shown in the Figure 12.4(b), which is coincided with the theoretical prediction. And the zero diffraction order located at the center of the diffraction pattern. We also fabricated the quasi-PCs structures via five-beam interference formed by RILS, experimental results related to the micro-structure image and far-field diffraction pattern are shown in Figs. 12.4(c) and (d).

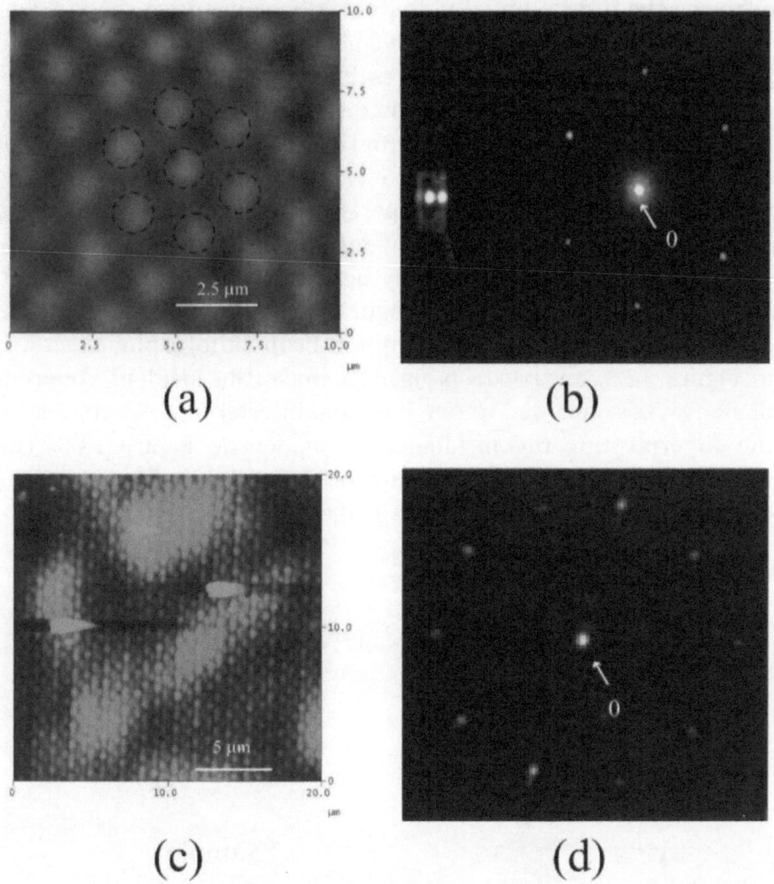

Figure 12.4 The scanned image by AFM of micro-structures fabricated by three beams (a) and five beams (c), and the corresponding far-field diffraction of the sample (b) and (d), respectively.

12.5 Further Analysis

Above all, we showed the capability of RILS theoretically and experimentally. PC/quasi-PC structures were fabricated by using this RILS system with three-beam interference and five-beam interference, respectively. The samples were based on the PDLC materials, which can be tuned electrically due to the electrically tunable property of the liquid crystal. Although this configuration achieved the interferential lithographic by converging the incident beams formed by DMD chip with convex lens to a point, it failed to generate the throughput structures due to its limited areas. First, the effective area of DMD chip is limited (about 1 × 1 cm). Therefore, it is difficult to achieve beamlets with large radius. Next, regardless of the radius of the beam, all of the incident beam would be generally focused due to the intrinsic property of convex lens. Although the focal point would be enlarged owing to the diffraction limitation, the interferential region can only be extended to about 1 mm. To address this issue, a novel configuration is proposed to fabricate large-area periodic structures by multi-beam lithography. As shown in Figure 12.5, an axicon is used to make the incident coherent plane waves converged to one position. Interference occurs within the superposition region labeled as shadow in Figure 12.5. The interference area will be determined by the apex angle α and the beam width r, which can be described as

$$L = r\left(1 - \frac{\tan\beta}{\tan\alpha}\right) \quad (12.3)$$

where α, β, r, and L refer to the labels in Figure 12.5.

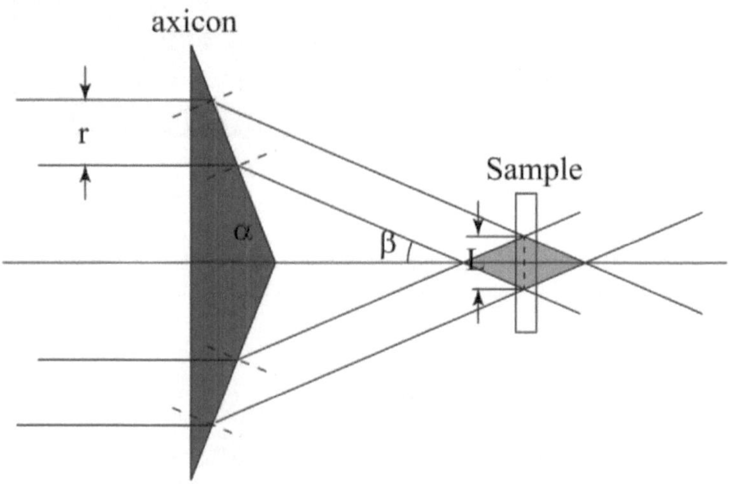

Figure 12.5 The novel configuration to achieve larger effective region by use of an axicon.

If a PDLC sample is placed in this position, electrically tunable PCs structures can be achieved with relative large effective area, at the same time, the azimuth angles of the incident beams can be adjusted by the DMD chip programmably. Nevertheless, the polar angles, which decide the optical lattice of the formed structures, cannot be tuned by changing the position of beamlets in the DMD chip as the aforementioned system with convex lens because the convergence angles of all parallel beams are equal each other for the axicon lens regardless of the beam position with respect to the optical axis.

12.6 Conclusion

In this chapter, we demonstrated the methods to achieve a reconfigurable interference lithographic system based on a DMD chip. This device can be used to fabricate PC/quasi-PC structures on photoresist with flexible capability to tune the parameters that determine the structure constants of PCs or quasi-PCs. We have made use of this RILS to fabricate electrically tunable PCs on PDLC materials successfully. Another approach that could achieve larger effective area lithography was also proposed by virtue of the combination of the axicon and the DMD device.

References

1. Deubel, M., Freymann, G. V., Wegener, M., Pereira, S., Busch, K., and Soukoulis, C. M. (2004). Direct laser writing of three-dimensional photonic-crystal templates for telecommunications, *Nat. Mater.* , 3, 444–447.
2. John, S. (1987). Strong localization of photons in certain disordered dielectric superlattices, *Phys. Rev. Lett.*, 58, 2486–2489.
3. Yablonovitch, E. (1987). Inhibited spontaneous emission in solid-state physics and electronics, *Phys. Rev. Lett.*, 58, 2059–2062.
4. Kim, D.H., Cho, C.O., Roh, Y.G., Jeon, H., Park, Y.S., Choi, W.J., and Park, Q.H. (2005). Enhanced light extraction from GaN-based light-emitting diodes with holographically generated two-dimensional photonic crystal patterns, *Appl. Phys. Lett.*, 87, 203508.
5. Guldin, S., Huttner, S., Kolle, M., Welland, M. E., Buschbaum, P. M., Friend, R. H., Steiner, U., and Tetreault, N. (2010). Dye-sensitized solar cell based on a three-dimensional photonic crystal, *Nano Lett.*, 10, 2303–2309.

6. Ho, K. M., Chan, C. T., Soukoulis, C. M., Biswas, R., and Sigalas, M. (1994). Photonic band gaps in three dimensions: new layer-by-layer periodic structures, *Solid State Commun.*, 89, 413–416.
7. Campbell, M., Sharp, D. N., Harrison, M. T., Denning, R. G., and Turberfield, A. J. (2000). Fabrication of photonic crystals for the visible spectrum by holographic lithography, *Nature*, 404, 53–56.
8. Yang, S., Megens, M., Aizenberg, J., Wiltzius, P., Chaikin, P. M., and Russel, W. B. (2002). Creating periodic three-dimensional structures by multibeam interference of visible laser, *Chem. Mater.*, 14, 2831–2833.
9. Miklyaev, Y. V., Meisel, D. C., Blanco, A., and Freymann, G. V. (2003). Three-dimensional face-centered-cubic photonic crystal templates by laser holography: fabrication, optical characterization, and band-structure calculations, *Appl. Phys. Lett.*, 82, 1284–1286.
10. Wu, L. J, Zhong, Y. C., Chan, C. T., and Wong, K. S. (2005). Fabrication of large area two- and three-dimensional polymer photonic crystals using single refracting prism holographic lithography, *Appl. Phys. Lett.*, 86, 241102.
11. Kondo, T., Matsuo, S., Juodkazis, S., and Misawa, H. (2001). Femtosecond laser interference technique with diffractive beam splitter for fabrication of three-dimensional photonic crystals, *Appl. Phys. Lett.*, 79, 725–727.
12. Zhang, X. S., Li, S. S., Liu, S., Lin, H., Chen, X. Y., and Ren, X. C. (2007). Fabrication of 2D photonic crystals with micron to sub-micron hexagonal lattices using single-exposure holographic technique, *Proc. SPIE*, 6832, 683205.
13. Liu, Y. J., and Sun, X. W. (2006). Electrically tunable two-dimensional holographic photonic crystal fabricated by a single diffractive element, *Appl. Phys. Lett.*, 89, 171101.
14. Hsieh, M. L., and Lan, Y. S. (2008). Controllable fabrication of the micropore shape of two-dimensional photonic crystals using holographic lithography, *J. Vac. Sci. Technol.* B, 26, 914–917.
15. Xavier, J., Boguslawski, M., Rose, P., Joseph, J., and Denz, C. (2010). Reconfigurable optically induced quasicrystallographic three-dimensional complex nonlinear photonic lattice structures, *Adv. Mater.*, 22, 356–360.
16. Xavier, J., Rose, P., Terhalle, B., Joseph, J., and Denz, C. (2009). Three-dimensional optically induced reconfigurable photorefractive nonlinear photonic lattices, *Opt. Lett.*, 34, 2625–2627.

17. Chen, K. M., Ren, H. W., and Wu, S. T. (2009). PDLC-based VOA with a small polarization dependent loss, *Opt. Comm.*, 282, 4374–4377.

18. Liu, Y. J., Sun, X. W., Elim, H. I., and Ji, W. (2006). Gain narrowing and random lasing from dye-doped polymer- dispersed liquid crystals with nanoscale liquid crystal droplets, *Appl. Phys. Lett.*, 89, 011111.

19. Burning, T. J., Natarajan, L. V., Tondiglia, V. P., and Sutherland, R. L. (2000). Holographic polymer-dispersed liquid crystals (H-PDLCS), *Annu. Rev. Mater. Sci.*, 30, 83–115.

Chapter 13

Silicon Waveguide-Based Polarization Rotator

Huijuan Zhang,[a,*] Jing Zhang,[b] Shiyi Chen,[a] Liang Ding,[a] Tsung-Yang Liow,[a] Mingbin Yu,[a] and Guo-Qiang Lo[a]

[a] *NanoPhotonics Program, Institute of Microelectronics, Agency for Science, Technology and Research (A*STAR), 1 Science Park Drive, Singapore Science Park I, 118221 Singapore*

[b] *Optical Metrology, National Metrology Centre, Agency for Science, Technology and Research (A*STAR), 1 Science Park Drive, Singapore Science Park I, 118221 Singapore*

*zhangh@ime.a-star.edu.sg

Silicon-based photonic circuit is very promising as a platform for high-speed communication due to its compatibility in the conventional CMOS process as well as the ability to integrate electronic and optical components on the same silicon chip. However, due to the high structural birefringence of silicon waveguides, which may cause unfavorable polarization-dependent performance, polarization transparent circuit is always preferred. A silicon waveguide-based polarization rotator is then a key device for polarization diversity circuit. In this chapter, we present two fabrication methods to obtain

Advances in Nanodevices and Nanofabrication: Selected Publications from Symposium of Nanodevices and Nanofabrication in ICMAT2011
Edited by Qing Zhang, W. I. Milne, and Jianping Zou

ISBN 978-981-4364-54-6 (Hardcover), 978-981-4364-55-3 (eBook)
www.panstanford.com

silicon waveguide-based polarization rotator. The performance of the polarization rotator is evaluated through finite-difference-time-domain simulations and also experiments. The experimental results show that an efficient 12.4 dB extinction ratio is achieved through a polarization rotator.

13.1 Introduction

Silicon photonics have received much attention in the recent years for its ability to integrate electronic and optical components on the same silicon chip.[1,2] Optical waveguides based on silicon (Si) and silica (SiO_2) material systems, i.e., Si waveguides, allow ultrasmall photonic circuits because of the strong refractive index contrast between Si (3.5) and SiO_2 (1.45).[3,4] Modulators,[5,6] photodetectors,[7,8] filters,[9,10] switches[11,12] and other passive devices[13,14] based on Si waveguides have been demonstrated. However, Si waveguide has large structural birefringence that causes polarization-mode dispersion, polarization-dependent loss, and polarization-dependent wavelength characteristics. In general, the polarization sensitivity could cause a polarization-dependent wavelength shift and phase shift in optical devices, such as spectrometers, Mach–Zehnder interferometers, and ring resonators, which are unfavorable and hence is preferred to be excluded. Therefore, polarization control and manipulation are crucial to the design and the operation of optical devices. Great effort has been spent on reducing the polarization sensitivity and eventually achieves polarization transparent optical circuit.[15–17] A polarization rotator then becomes a key element in the polarization diversity circuit. The light with arbitrary polarization from the source will be first split into its orthogonal components (TE and TM) by a polarization splitter. By further rotating one of the components into its perpendicular direction using a polarization rotator, the two components having same polarizations may be processed individually through identical paths to achieve polarization transparent response.

Here we report a Si waveguide-based polarization rotator using a silicon-on-insulator (SOI) platform. Two fabrication methods are proposed and discussed. The device performance is evaluated through simulation and investigated through experiments.

13.2 Design and Simulation

A schematic diagram of the polarization rotator is shown in Figure 13.1. The device consists of a pair of Si channel waveguides having dimensions of 400 nm (width) × 200 nm (height) at the input port (named as horizontal waveguide) and 200 nm (width) × 400 nm (height) at the output port (named as vertical waveguide). The mode-evolution-based rotation occurs within the transition region, which allows gradual transition from horizontal waveguide to vertical waveguide by a tapered structure. The cross-sectional views at the horizontal waveguide, transition region and the vertical waveguide are shown in Figure 13.1 as indicated by the dotted lines AA′, BB′ and CC′. The working principle of rotators based on silicon nitride waveguides having dimension of 800 nm (width) × 400 nm (height) was first described by Watts and Haus in [20]. The efficient rotation with low insertion loss is achieved by matching the effective refractive index of input light mode in the horizontal waveguide (AA′) and that of output light mode in the vertical waveguide (CC′), and vice versa.

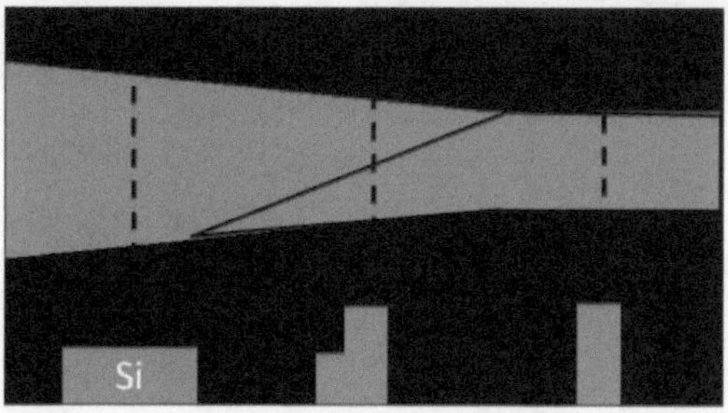

Figure 13.1 Schematic of a polarization rotator. The insets are cross-sectional views of the rotator as indicated by the dotted lines AA′, BB′ and CC′, respectively.

The performance of the rotator is studied by using three-dimensional finite-difference-time-domain (FDTD) simulation. The operating wavelength is 1.55 μm. The waveguide core is Si with an index of 3.5, while the cladding is SiO_2 with an index of 1.45. To simulate the polarization rotation effect, TE and TM modes were launched at the input plane $z = 0$ μm and outputs at the end of the rotator were monitored. E_x is the electrical filed in x-axis and E_y is the electrical field in y-axis.

Figure 13.2 shows the light propagation along a rotator having 20 μm transition length when only launching TE at the input, corresponding to E_x component. Figures 13.2(a) and (b) show E_x and E_y components in the rotator and the mode profiles at the input and output, respectively. It is observed that E_x component gradually decreases and E_y component appears at the initial portion of the transition region. At the end of the transition region, E_y dominates and E_x disappears. The effective refractive index of TE and TM modes at the input and output waveguide is matched and calculated to be 2.26 at 1550 nm wavelength. Hence, the electrical field vector rotated. The power extinction ratio (PER), i.e., the power of TM mode versus that of TE mode is calculated to be 27.4 dB, indicating almost complete rotation.

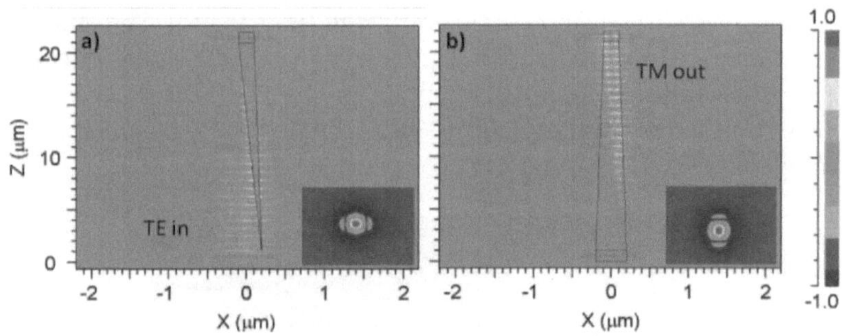

Figure 13.2 (a) E_x component propagation along the polarization rotator with transition length 20 μm and 0 nm tip; the inset is TE-mode profile in the input waveguide; (b) E_y component propagation along the polarization rotator; the inset is TM-mode profile in the output waveguide.

Similarly, to observe the rotation effect from TM to TE, only TM mode is launched at the input waveguide. Figure 13.3 shows E_x and E_y components in the rotator and the mode profiles at the input and output, respectively. The effective refractive index of TM mode at the input port is 1.70, while the effective refractive index of TE mode at the output port is 1.69 that to match with the input mode. Polarization rotation also occurs as it is seen that E_y disappears and E_x dominates at the end of transition region. PER of TE power versus TM power is calculated to be 14.9 dB.

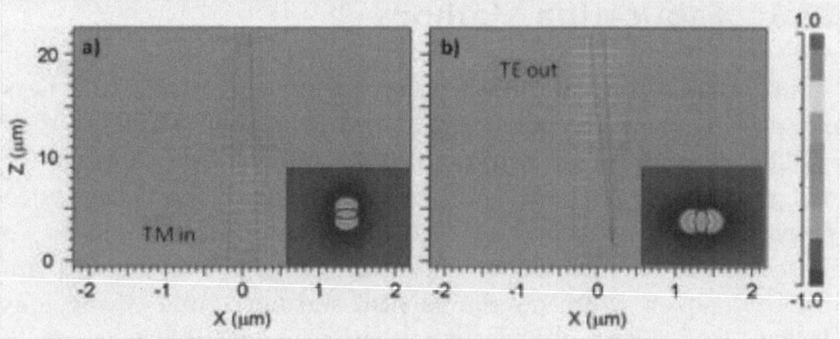

Figure 13.3 (a) E_y component propagation along the polarization rotator with transition length 20 µm and 0 nm tip; the inset is TM-mode profile in the input waveguide; (b) E_x component propagation along the polarization rotator; the inset is TE-mode profile in the output waveguide.

Simulation using the same structure in the reverse direction is also carried out and the results show that similar rotation effect is achieved. In addition, devices with various transition lengths, (10 µm, 20 µm and 40 µm) were studied. The respective PER results are shown in Table 13.1. The simulations presented show the effective polarization rotation by the structure proposed under both launching condition. TE mode in 400 nm (width) × 200 nm (height) waveguide rotated to TM mode in 200 nm (width) × 400 nm (height) waveguide. On the other hand, TM mode is also rotated to TE mode in the same structure. In general, the PER is better when the transition length increases as the mode evolution happens in a more gradual manner. Moreover, it is noted that TE-TM rotation has a better PER than TM-TE rotation in the obverse direction, while TM-TE rotation has a better PER than TE-TM rotation in the obverse direction.

Table 13.1 Simulation results for power extinction ratio of the polarization rotators

Polarization rotator transition length (µm)	TE-TM obverse PER (dB)	TM-TE obverse PER (dB)	TE-TM reverse PER (dB)	TM-TE reverse PER (dB)
10	14.45	1.71	1.33	14.55
20	27.76	14.78	14.94	28.04
40	32.66	19.37	19.97	32.83

13.3 Fabrication Methods

The process starts with a SOI wafer having active layer thickness of 340 nm. Blanket epitaxy growth of Si is then carried out to thicken the active layer thickness to the desired value 400 nm for the aforementioned vertical waveguide height. Then waveguide structures are patterned by optical lithography and etched to buried oxide surface (BOX). The waveguide now have 200 nm and 400 nm in width for the vertical and horizontal waveguides design. Next, the wafer is patterned again by optical lithography to selectively protect vertical waveguide (VWG) and also the transition region by photoresist and expose the horizontal waveguide region (HWG). Careful etching is then carried out to thin down the exposed waveguide by 200 nm Si etch.

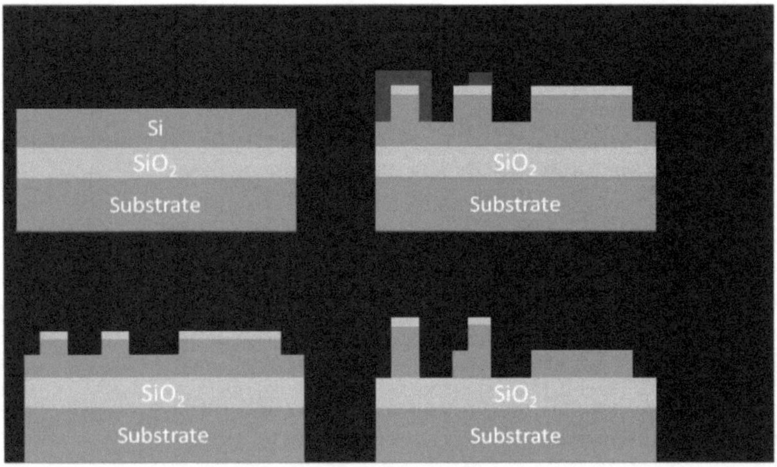

Figure 13.4 (a) Blanket epitaxy (EPI) growth to thicken Si layer to 400 nm. (b) Waveguide pattern and partial etch. (c) Second time pattern to selectively expose the waveguides. (d) Fully Si etch to form waveguide structures.

An alternative method is illustrated in Figure 13.4. The major steps are similar except the use of a hardmask layer (SiO_2). The wafer goes through epitaxy growth, and a thin SiO_2 layer is deposited as the hardmask. The wafer is then lithographical patterned to define the waveguide structures. As shown in Figure 13.4(b), instead of etching Si till BOX surface, Si layer is partially etched (200 nm). Next, the waveguides with hardmask is patterned with second lithography to expose the horizontal waveguide region (Figure 13.4(c)). The hardmask is then etched to transfer the lithographical pattern followed by Si etching to BOX surface. (Figure 13.4(d)) In this way, the remaining 200 nm Si layer from first time etching is thinned down together with the horizontal waveguide region, which serves as an indicator of etching thickness for accurate process control.

13.4 Results and Discussion

Figure 13.5 shows the SEM images of the polarization rotator obtained after the fabrication. Figure 13.5(a) shows the transition region between the horizontal waveguide (bottom) and the vertical waveguide (top). Figures 13.5(b) and 13.6(c) are the SEM images of the waveguide obtained using the first fabrication method presented. A thick white coating is observed on the sidewall along the waveguide. A schematic drawing is inserted in Figure 13.5(b) to illustrate the structure obtained. The wafer was cleaned by diluted hydrofluoric acid (DHF) for a few times to attempt to get away the coating. However, the material was not dissolved in DHF and remained. The oxide loss indicated in Figure 13.5(b) is the result of DHF clean. Local inspection of the waveguide is done by TEM and the result is shown in Figure 13.5(c). As observed, the coating has a thickness of about 13 nm and is even 47 nm higher than the waveguide surface. The composition of the coating is not clear yet. We suspect that the coating is non-volatile polymer resulting from long-time plasma bombardment on BOX surface during the second time etching of Si horizontal waveguide. Figures 13.5(d) and 13.5(e) show the SEM images of the waveguide obtained using the second fabrication method. It is clearly observed that the sidewall of the waveguide obtained is very clean and free of any polymer coating. This is because of non-exposed BOX surface during the second time etching. As presented in the fabrication section, the key difference of the two proposed method is the Si etching sequence: fully etched Si followed by partial etching versus partial etching followed by thinning down together to obtain fully etched waveguide. The second method is seen to be more advantageous as the device obtained is free of polymer coating, which may affect the device performance and also has better process stability. The second method allows better controlled etching thickness because the wafer-scale etching area enables large etching tolerance during the second step etching. In contrast, a slight over-etching in the first method when partial etching of slim waveguide may result in undesired large variation of the device structure. In addition, the devices obtained by the first method have a much higher propagation loss, close to 10 dB, for both TE and TM mode than the devices obtained by the second method. This high propagation should be resulting from the coating that we observed in SEM images, which should definitely be avoided.

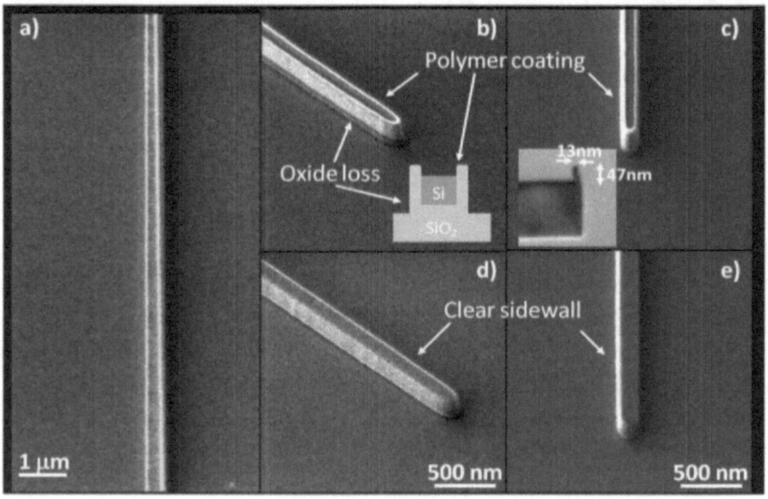

Figure 13.5 Scanning microscope images of (a) transition region between the horizontal waveguide and the vertical waveguide. (b) Waveguide obtained using the first fabrication method. (c) The waveguide in (b) tilted 45°. (d) Waveguide obtained using the second fabrication method. (e) The waveguide in (d) tilted 45°. The inset in (b) is the schematic of the structure obtained. The inset in (c) is the TEM image of the waveguide with coating having thickness of 13 nm and 47 nm height above the waveguide surface.

Table 13.2 Extinction ratio of the polarization rotators

Transition length (μm)	TE-TM PER (dB)	TM-TE PER (dB)
40	4.3	6.3
60	8.8	9.7
80	7.9	12.4
100	9.4	8.6

The device performance is characterized experimentally. A tunable laser source at wavelength 1550 nm was used in the experiments. The polarization of the light was controlled by a polarizer and polarization maintaining fibers were used throughout the tests. Light was coupled into waveguide through invert tapper structure having tip size of 180 nm.[26] After light propagating along the polarization rotator, output light was coupled back into PM fiber. At the output end, another polarizer was connected to measure the power of TE and TM components separately to examine the polarization rotation effect. The experiments were done under TE and TM launching conditions while using the device in both obverse and reverse directions. The transition lengths of the rotators are from 40 to 100 μm in 20 μm intervals. Table 13.2 shows the power extinction ratios (PER) of the polarization rotators. Efficient rotation between TE and TM modes is achieved: 9.4 dB PER from TE to TM and 12.4 dB PER from TM to TE. In general, PER increases as the transition length increases as observed in Table 13.1. The increment of PER is not proportional with the increment of the transition length as the complete rotation may occur above certain transition length, in our case longer than 60 μm is sufficient. We observed polarization rotation effect in all the measurements using the fabricated device. However, the rotator performance in terms of PER is better for TE-to-TM rotation in the direction of horizontal waveguide to vertical waveguide and for TM-to-TE rotation in the direction of vertical waveguide to horizontal waveguide. This observation should be reasonable given the fundamental preferred mode in each waveguide structure. In the horizontal waveguide, TE mode is confined well in the core of the channel waveguide. On the other hand, TM mode is better confined in the vertical waveguide. The mode profiles have been presented in Figures 13.2 and 13.3. From these mode profiles, we believe the propagation loss of TE mode in vertical waveguide and TM mode in horizontal waveguide is intrinsically higher than TM mode in vertical waveguide and TE mode in horizontal waveguide, respectively. Hence, even the rotation effect of the TM-to-TE rotation in the direction of horizontal waveguide to vertical waveguide and for the TE-to-TM rotation in the direction of vertical waveguide to horizontal waveguide is as good as those aforementioned obverse directions, the high propagation loss in the waveguide would strongly affect the PER and hence the device performance. The experimental results showing the preferred rotation direction is in line with simulation results as observed in Table 13.1.

13.5 Conclusions

In conclusion, we have presented two methods to fabricate a Si waveguide-based polarization rotator using CMOS compatible process. The devices obtained were characterized for the performance under different launching conditions. Efficient polarization rotation of 9.4 from TE to TM and 12.4 dB from TM to TE extinction ratio is achieved. With this polarization rotator as the essential building block, future new designs of polarization transparent circuit will be very promising.

References

1. T. Tsuchizawa, K. Yamada, H. Fukuda, T. Watanabe, J. Takahashi, M. Takahashi, T. Shoji, E. Tamechika, S. Itabashi, H. Morita, Microphotonics devices based on silicon micro-fabrication technology, *IEEE J. Sel. Top. Quantum Electron.*, **11**, 232–240 (2005).
2. D. A. B. Miller, Optical interconnects to silicon, *IEEE J. Sel. Top. Quantum Electron.*, **6**, 1312–1317 (2000).
3. S. Nakamura, T. Chu, M. Ishizaka, M. Tokushima, Y. Urino, M. Sakauchi, I. Nishioka, K. Fukuchi, Ultra-small one-chip color-less multiplexer/demultiplexer using silicon photonic circuit, in *Proceedings of the ECOC*, 2008, pp. 175–176.
4. F. Xia, L. Sekaric, Y. Vlasov, Ultracompact optical buffers on a silicon chip, *Nat. Photon.*, **1**, 65–71 (2007).
5. A. Liu, R. Jones, L. Liao, D. Samara-Rubio, D. Rubin, O. Cohen, R. Nicolaescu, M. Paniccia, A high-speed silicon optical modulator based on a metal-oxide-semiconductor capacitor, *Nature*, **427**, 615–618 (2004).
6. Q. Xu, S. Manipatruni, B. Schmidt, J. Shakya, M. Lipson, 12.5 Gbit/s carrier-injection-based silicon micro-ring silicon modulator, *Opt. Express*, **15**, 430–436 (2007).
7. T.-Y. Liow, K.-W. Ang, Q. Fang, J.-F. Song, Y.-Z. Xiong, M.-B. Yu, G.-Q. Lo, Silicon modulators and germanium photodetector on SOI: monolithic integration, compatibility, and performance optimization, *IEEE J. Sel. Top. Quantum Electron.*, **16**, 307–315 (2010)
8. L. Chen, P. Dong, M. Lipson, High performance germanium photodetectors integrated on submicron silicon waveguides by low temperature wafer bonding, *Opt. Express*, **16**, 11513–11518 (2008).

9. P. Dong, W. Qian, H. Liang, R. Shafiiha, X. Wang, D. Feng, G. Li, J. E. Cunningham, A. V. Krishnamoorthy, M. Asghari, 1 × 4 reconfigurable demultiplexing filter based on free-standing silicon racetrack resonators, *Opt. Express*, **18**, 24504–24509 (2010).

10. M.S. Rasras, D. M. Gill, M. P. Earnshaw, C. R. Doerr, J. S. Weiner, C. A. Bolle, Y.-K. Chen, CMOS silicon receiver integrated with Ge detector and reconfigurable optical filter, *IEEE Photon. Technol. Lett.*, **22**, 112–114 (2010).

11. Y. Shoji, K. Kintaka, S. Suda, H. Kawashima, T. Hasama, H. Ishikawa, Low-crosstalk 2 × 2 thermo-optic switch with silicon wire waveguides, *Opt. Express*, **18**, 9071–9075 (2010).

12. M. Yang, W. M. J. Green, S. Assefa, J. Van Campenhout, B. G. Lee, C. V. Jahnes, F. E. Doany, C. L. Schow, J. A. Kash, Y. A. Vlasov, Non-blocking 4 × 4 electro-optic silicon switch for on-chip photonic networks, *Opt. Express*, **19**, 47–54 (2010).

13. H. Yamada, T. Chu, S. Ishida, Y. Arakawa, Optical directional coupler based on Si-wire waveguides, *IEEE Photon. Technol. Lett.*, **17**, 585–587 (2005).

14. B. Guha, B. B. C. Kyotoku, M. Lipson, CMOS-compatible athermal silicon microring resonators, *Opt. Express*, **18**, 3487–3493 (2010).

15. T. Barwicz, M. R. Watts, M. A. Popovic, P. T. Rakich, L. Socci, F. X. Kartner, E. P. Ippen, Henry. I. Smith, Polarization-transparent microphotonic devices in the strong confinement limit, *Nat. Photon.*, **1**, 57–60 (2007).

16. H. Fukuda, K. Yamada, T. Tsuchizawa, T. Watanabe, H. Shinojima, S. Itabashi, Silicon photonic circuit with polarization diversity, *Opt. Express*, **16**, 4872–4880 (2008).

17. J. Zhang, H. Zhang, S. Chen, M. Yu, G. Q. Lo, D. L. Kwong, A tunable polarization diversity silicon photonics filter, *Opt. Express*, Accepted.

18. H. Fukuda, K. Yamada, T. Tsuchizawa, T. Watanabe, H. Shinojima, S. Itabashi, Ultrasmall polarization splitter based on silicon wire waveguides, *Opt. Express*, **14**, 12401–12408 (2006).

19. W. N. Ye, D.-X. Xu, S. Janz, P. Waldron, P. Cheben, N. G. Tarr, Passive broadband silicon-on-insulator polarization splitter, *Opt. Lett.*, **32**, 1492–1494 (2007).

20. M. R. Watts, H. A. Haus, Integrated mode-evolution-based polarization rotators, *Opt. Lett.*, **30**, 138–140 (2005).

21. J. Zhang, M. Yu, G.-Q. Lo, D.-L. Kwong, Silicon-waveguide-based mode evolution polarization rotator, *IEEE J. Sel. Top. Quantum Electron.*, **16**, 53–60 (2010).

22. Z. Wang, D. Dai, Ultrasmall Si-nanowire-based polarization rotator, *J. Opt. Soc. Am. B*, **25**, 747–753 (2008).

23. L. Chen, C. R. Doerr, Y.-K. Chen, Compact polarization rotator on silicon for polarization-diversified circuits, *Opt. Lett.*, **36**, 469–471 (2011).

24. H. Fukuda, K. Yamada, T. Tsuchizawa, T. Watanabe, H. Shinojima, S.-I. Itabashi, Polarization rotator based on silicon wire waveguides, *Opt. Express*, **16**, 2628–2635 (2008).

25. Y. Shani, R. Alferness, T. Koch, U. Koren, M. Oron, B. I. Miller, M. G. Young, Polarization rotation in asymmetric periodic loaded rib waveguides, *Appl. Phys. Lett.*, **59**, 1278–1280 (1991).

26. V. R. Almeida, R. R. Panepucci, M. Lipson, Nanotaper for compact mode conversion, *Opt. Lett.*, **28**, 1302–1304 (2003).

Chapter 14

Silicon Microring Resonator-Based Thermo-Optic Wavelength Selective Switch for Optical Communication Network

Prabhathan Patinharekandy,[a,b] Jing Zhang,[c] Huijuan Zhang,[b] Guo-Qiang Lo,[b] and Murukeshan Vadakke Matham[a,*]

[a] *School of Mechanical and Aerospace Engineering, Nanyang Technological University, 50 Nanyang Avenue, 639798 Singapore*

[b] *Institute of Microelectronics, Agency for Science, Technology and Research (A*STAR), 11 Science Park Road, 117685 Singapore*

[c] *Optical Metrology, National Metrology Centre, Agency for Science, Technology and Research (A*STAR), 1 Science Park Drive, Singapore Science Park I, 118221 Singapore*

*mmurukeshan@ntu.edu.sg

Optical wavelength selective switches (WSS) are essential devices for high-capacity optical network systems based on reconfigurable optical add-drop multiplexing (ROADM) and/or optical cross connect (OXC). They can dynamically route, block and attenuate all dense wavelength division multiplexed (DWDM) wavelengths within a network node. The switches used in such systems must be compact

Advances in Nanodevices and Nanofabrication: Selected Publications from Symposium of Nanodevices and Nanofabrication in ICMAT2011
Edited by Qing Zhang, W. I. Milne, and Jianping Zou

ISBN 978-981-4364-54-6 (Hardcover), 978-981-4364-55-3 (eBook)
www.panstanford.com

and inexpensive, have low power consumption, and have the required design flexibility. Here, the design, fabrication and experimental characterization of a thermo-optic (TO) WSS on a silicon-on-insulator (SOI) platform is presented with wide wavelength tunability, low thermal cross talk and spectral characteristics suitable for DWDM optical communication network.

14.1 Introduction

Many of the photonic interconnect network architectures and wavelength filtering devices rely on the microring resonators (MRR) as their WSS because of their versatility, compact size and energy efficiency [1–3]. Mainly two approaches are employed in MRR-based devices to tune its resonant wavelength, using its electro-optic (EO) or TO effect. TO method in silicon has been shown to be highly efficient in circuit switched optical networks, providing long-term stability, wide wavelength tunability and with an operating time scale of a few μs under direct heating [4, 5]. Also, when compared to EO tuning, the TO tuning provides an easy fabrication procedure and there is no inherent loss present in the method. For a WSS in optical communication network, hitless wavelength switching and tuning is highly desirable, as this will avoid the disturbances on the other wavelength channels and allow them to pass through the device. One of the methods to achieve hitless tuning in MRR WSS is to use series coupled ring resonators through Vernier or non Vernier configurations. When the resonance wavelengths match inside the rings, the device drops that wavelength to make the switch into ON state. When the resonance mismatch occurs through any applied cavity detuning, the device goes to OFF state and allow the channel to pass through the device undisturbed. Hitless TO-WSS has been demonstrated in polymer, dielectric (SiO_2) and SiO_2–Si_3N_4 ring resonators through Vernier configuration [6–8]. While Vernier configuration is an easy way to expand the free spectral range (FSR) of the device, the suppressed channels undergo unnecessary dispersion at the through port, causing bit errors to the through port channels. Also the directly coupled rings will introduce thermal cross talk between the rings causing thermal interference on other wavelength channels. Thus, to achieve a truly hitless configuration in a TO-double ring WSS, Vernier effect and the thermal interference on the through port channels should be avoided. At the same time, WSS should allow the selection and routing of a single wavelength channel from a broadband (~C + L band) input, with narrow channel spacing and low power consumption.

Here a SOI microring resonator-based WSS is demonstrated with two stage filtering configuration, so that thermal cross talk between the rings is considerably reduced. A small difference in the radius of the rings will allow selection and routing of a single wavelength channel at the drop port with a wide FSR (~95 nm). The WSS is characterized for its spectral behavior. Its wavelength selectivity is demonstrated through discrete and fine wavelength tuning with low power consumption.

14.2 Device Design and Fabrication

The device design is shown in Figure 14.1(a). It has two rings coupled serially through an intermediate waveguide with a separation of 20 μm between them. The ring with higher radius is the first ring (R1) and ring with smaller radius is the second ring (R2). The FSR is the span of wavelength between two adjacent resonances of a single ring. Here the FSR of the R1 is termed as FSR_1 and FSR of R2 is termed as FSR_2. Only the wavelengths that are in resonance with both rings will be dropped from the device with a separation between them given by extended FSR ($FSR_{extended}$) at the drop port.

The Vernier principle is expressed by the expression

$$FSR_{extended} = m_1 FSR_1 = m_2 FSR_2 \quad (14.1)$$

Where m_1 and m_2 are co-prime integers and is known as FSR expansion factor. To get the maximum FSR expansion, the relationship between m_2 and m_1 is set as follows:

$$m_2 = m_1 - 1 \quad (14.2)$$

To satisfy this condition and to get an FSR expansion of ~95 nm, the ring radii are set as R_1 = 10 μm and R_2 = 9 μm.

With these ring radius values, the device fabrication is done on an 8 inch SOI wafer with a 220 nm top silicon layer thickness and 2 μm buried oxide layer. The waveguide structure is patterned with mask-based deep ultra-violet (DUV) lithography. Silicon etching is done with inductively coupled plasma (ICP) reactive ion etching (RIE). To increase the coupling efficiency between input fiber and the device, the waveguide ends are terminated with spot size converters (SSC) having a length of 200 μm and tip width of 180 nm. Silicon dioxide (SiO_2) cladding layer of 2 μm is deposited over the silicon waveguide layer through plasma-enhanced chemical vapor deposition (PECVD) process. After that the titanium (Ti) microheaters are fabricated over the structure through DUV lithography and dry etching. The heater elements are 100 nm thick 800 nm wide. After this, another 1 μm SiO_2 cladding layer is deposited and contact holes are etched. Afterward, aluminum (Al) wire and connection pads are fabricated on top of structure for external connection. Figure 14.1(b) shows the cross-sectional image of the SOI-based device showing various dimensions. The cross-sectional dimensions of bus waveguide and ring waveguide are 470 × 220 nm and 525 × 220 nm, respectively. The coupling gaps are 200 nm for all the four bus-ring couplers. Different SEM and optical microscope images of the device are shown in Figure 14.2.

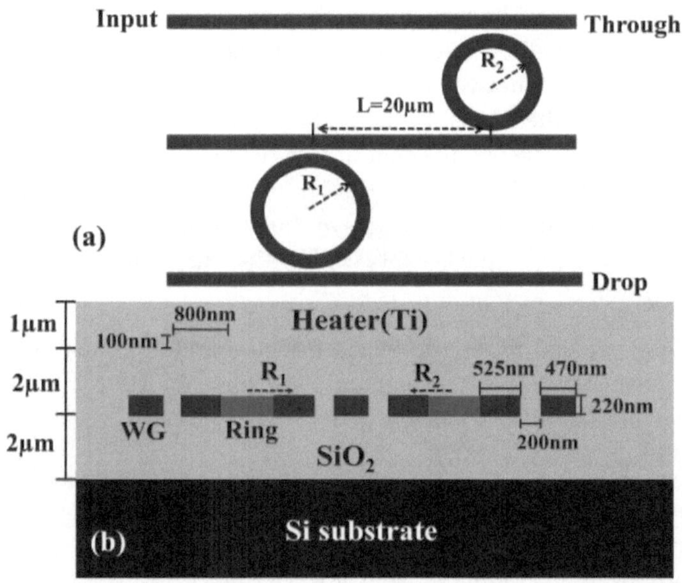

Figure 14.1 (a) Schematic diagram of the double ring WSS (b) Cross-sectional view of the device, showing waveguides and microheaters.

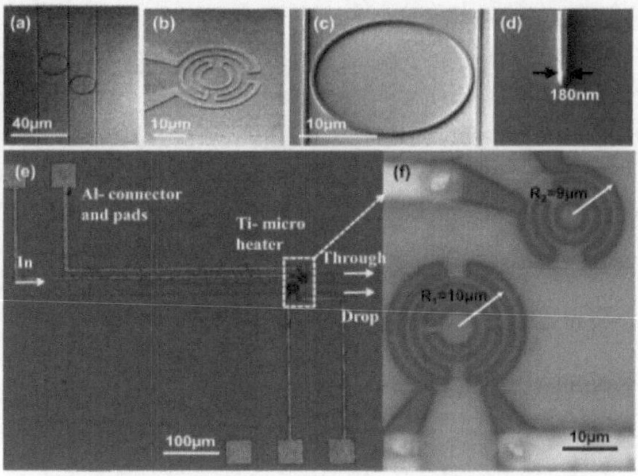

Figure 14.2 SEM images of (a) double ring WSS, (b) Ti microheater, (c) single ring of radius 10 μm and (d) SSC tip. Optical microscope image of the (e) device and (f) magnified image of Ti-microheater and rings.

14.3 Experimental Results and Discussion

The device characterization is done using an amplified spontaneous emission (ASE) light source (EXFO FLS 2300B) with spectral range of 1530–1600 nm and optical spectrum analyzer (AQ6317B). A polarization controller (PC) is used to select a TE mode (electric field parallel to the substrate plane) from the ASE source. One *I–V* sweep and a voltage-current source meters are used to supply DC current to both microheaters.

14.3.1 Discrete and Fine Wavelength Tuning

Initially, with no electric power given to the rings, no drop wavelength is observed due to the resonance mismatch in the rings. The electrical power on R1 (P_1) is increased slowly while keeping the power on R2 (P_2) to 0 mW. The first peak is observed at 1554.28 nm with P_1 = 4.93 mW. When P_1 is increased further to 5.73 mW, the device has gone to the OFF state. With a linear increase in P_1, the device has shown discrete wavelength switching from one drop wavelength to the other with equal wavelength spacing. The wavelength spacing is observed to be exactly the FSR of R2 (FSR_2), ~10.7 nm. Figure 14.3(a) shows the superimposed wavelength switching spectrum with increase in P_1. An average switching power of 7.3 mW is observed to switch from one drop wavelength to the other. The spectrum exhibited almost equal bandwidth, extinction and transmission power. The slight increase in the wavelength shift for higher powers is due to the group index variation with respect to the increase in wavelength. This is observed as a slight increase in the FSR of a ring resonance with increase in wavelength.

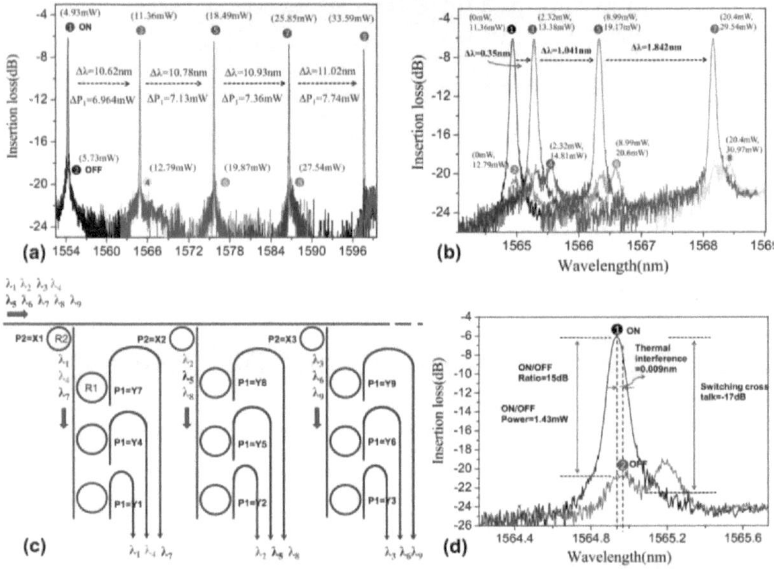

Figure 14.3 (a) Discrete wavelength tuning in WSS through electrical power variation (ΔP_1) applied on R1 (with P_2 = 0 mW) (b) Fine wavelength tuning in WSS through electrical power variations (ΔP_2, ΔP_1) applied on R2 and R1 (c) Schematic diagram showing the matrix array of the proposed WSS to select and route DWDM channels with different electrical powers (P_1, P_2) on the switches. (d) ON/OFF spectrum showing ON/OFF ratio, ON/OFF power, thermal interference and switching cross talk.

Wavelength fine-tuning is done through electrical power variation given to both heaters. The voltage on heater R2 is varied in steps of 2 V to set different electrical powers on R2, (P_2) from 0 mW to 20.4 mW. The power on R1 (P_1) is increased at each stage to catch the shifted resonance, so that the dropped spectrum is tuned in small wavelength ranges. Figure 14.3(b) shows the tuned drop spectrum from 1564.9 nm to 1568.2 nm with different electric powers (P_2, P_1) applied on the rings. The spectral shift has shown a linear variation with respect to the input electrical powers.

14.3.2 Wavelength Selection in a Broadband DWDM System

The device can drop channel wavelengths with a spectral spacing of 100 GHz and 50 GHz with a channel cross talk less than –15 dB, through the combined fine and discrete tunability. The average power difference (ΔP_2) applied on R2 to fine-tune the drop spectrum to 100 GHz and 50 GHz channel spacing is observed to be 5 mW and 2.5 mW, respectively. The WSS can be employed in a matrix array configuration as shown in the Figure 14.3(c) to demultiplex, select or route different DWDM wavelength channels in an optical communication network. The device can be reconfigured to any of the desired channels through electrical power variations applied on rings. Thermal cross talk between the rings will introduce undesirable effects on the switching characteristics of the device, in the form of thermal interference during switching. Here the thermal interference during switching is observed to be 0.009 nm with an applied ON/OFF power of 1.43 mW on R1 [see Figure 14.3(d)]. To estimate the amount thermal cross talk between the rings, spectral shift observations are made with varying electrical powers on the rings. Here, the thermal cross talk between R1 and R2 is defined as the amount of spectral shift in one ring when electrical power is varied in the other ring. The thermal cross talk is observed to be considerably smaller (0.85 GHz/mW, 4.2%) with the designed ring separation of 20 μm in the device.

In conclusion, SOI MRR-based WSS is experimentally demonstrated with two-stage filtering configuration, so that thermal cross talk between the rings is considerably reduced. The WSS is characterized for its spectral behavior and its wavelength selectivity is demonstrated using discrete and fine wavelength tuning with low power consumption. An average tuning powers of, 7.3 mW/FSR and 6.2 mW/nm is observed for its discrete and fine wavelength tunability, respectively. A matrix array configuration is proposed using the designed WSS to dynamically select and route DWDM wavelength channels in an optical communication system.

References

1. Johnnie. C., Noam. O., Caroline. P. L., Aleksandr. B., Hugo. L. R. L., Michal. L., Keren. B. (2011) Data transmission using wavelength-selective spatial routing for photonic interconnection networks, in *Optical Fiber Communication Conference*, p. OThQ3.
2. Goebuchi. Y., Kato. T., Kokubun. Y. (2007) Multiwavelength and multiport hitless wavelength-selective switch using series-coupled microring resonators, *IEEE. Photon. Technol. Lett.*, **19**, 671–673.
3. Shih-Jung. C., Chi-Yu. N., Zhipeng. W., Yung-Jui. C. (2008) A compact and low power consumption optical switch based on microrings, *IEEE Photon. Technol. Lett.*, **20**, 1021–1023.
4. Dong, P., Qian, W., Liang, H., Shafiiha, R., Feng, N.-N., Feng, D., Zheng, X., Krishnamoorthy, A. V., Asghari, M. (2010) Low power and compact reconfigurable multiplexing devices based on silicon microring resonators, *Opt. Express*, **18**, 9852–9858.
5. Sherwood-Droz, N., Wang, H., Chen, L., Lee, B. G., Biberman, A., Bergman, K., Lipson, M. (2008) Optical 4 × 4 hitless silicon router for optical networks-on-chip (NoC), *Opt. Express*, **16**, 15915–15922.
6. Goebuchi, Y., Kato, T., Kokubun, Y. (2006) Fast and stable wavelength-selective switch using double-series coupled dielectric microring resonator, *IEEE. Photon. Technol. Lett.*, **18**, 538–540.
7. Yamagata, S., Kato, T., Kokubun, Y. (2005) Non-blocking wavelength channel switch using TO effect of doubles series coupled microring resonator, *Electron. Lett.*, **41**, 593–595.
8. Geuzebroek, D., Klein, E., Kelderman, H., Baker, N., Driessen, A. (2005) Compact wavelength-selective switch for gigabit filtering in access networks, *IEEE Photon. Technol. Lett.*, **17**, 336–338.

Chapter 15

Tuning Plasmonic Resonance Wavelength by Ordered Silver Hole Arrays on Si and SiO_2

Hongwei Gao,[a] Benzhong Wang,[b] Jinghua Teng,[b] Soo Jin Chua,[b] and Ning Xiang[a,*]

[a] *Department of Electrical and Computer Engineering, National University of Singapore, 117576 Singapore*

[b] *Institute of Materials Research and Engineering, 117602 Singapore*

*elexn@nus.edu.sg

This chapter studies the tuning of plasmonic resonance wavelength by changing the hole diameter of an ordered silver hole array on silicon. The effect of the surrounding medium on the plasmonic resonance wavelength is also studied. It is observed that the plasmonic resonance wavelength is sensitive to the changing of silver holes' diameter. Our experiment shows an 11 nm red-shift in wavelength when the holes' diameter decreased by 60 nm. Simulation also shows that by further decreasing the Ag holes' diameter by 230 nm, 100 nm wavelength tuning can be achieved. Besides the method of adjusting the holes' diameter, wavelength tuning can also be achieved by changing the surrounding medium. Our experiment shows that

Advances in Nanodevices and Nanofabrication: Selected Publications from Symposium of Nanodevices and Nanofabrication in ICMAT2011
Edited by Qing Zhang, W. I. Milne, and Jianping Zou

ISBN 978-981-4364-54-6 (Hardcover), 978-981-4364-55-3 (eBook)
www.panstanford.com

for the same structure attached on Si and SiO_2 surfaces respectively, the plasmonic resonance wavelength is red-shifted for about 77 nm when the surface beneath the Ag nanoholes changes from Si to SiO_2.

15.1 Introduction

Plasmonic nanostructures have received considerable attention since their extraordinary optical transmission at some particular wavelength has been first announced in 1998 [1]. This particular wavelength corresponds to the maximum extinction (also the minimum reflection) of light from the plasmonic nanostructure, which highly relies on the size, shape and dielectric properties of the metal and its surrounding medium. The extraordinary optical phenomenon, the surface plasmon resonance, is due to the collective oscillation between electrons in metal and photons from light [2–3], which has enabled new fundamental science and found extensive applications. Many nanofabrication techniques, such as electron beam lithography (EBL), nano-imprint lithography (NIL) and focused ion beam (FIB), have been employed to create nanostructures with high-resolutions [4–6]. However, these techniques have inherent limitations of high cost, time-consuming nature and small area. Nanosphere lithography has shown to be a promising simple technique to fabricate large area nanostructures with low cost and much less time consumption compared with the techniques mentioned above.

Plasmonic nanostructures are promising candidates in the field of medical and biological sensors. In order to identify different elements, tuning the extinction wavelength is a critical requirement. Previous work using nanosphere lithography technique to tune plasmonic resonance wavelength mostly focused on metal nanoparticles. Traci R. Jensen successfully tuned the extinction wavelength of Ag nanoparticle arrays throughout the visible and near-infrared region by varying the nanoparticle parameters using nanosphere lithography technique [7]. Amanda J. Haes reported the plasmon resonance shift due to the changing of metal particle size and dielectric constant of metal [8]. It is also shown that the optical response of nanoparticle arrays is largely independent of interparticle separation, while the response from nanohole arrays exhibits a marked dependence on interhole separations [9].

Renlong Zhou reported the plasmonic resonance wavelength red-shifted while nanohole period increases with fixed hole diameter [10]. In this chapter, we study the tuning of plasmonic resonance wavelength based on Ag nanohole arrays with a fixed period but different diameters so as to change the interhole separations. The nanohole arrays are fabricated by nanosphere lithography method on Si substrate. We also study the effect of the surrounding medium to the plasmonic resonant wavelength by attaching the same Ag hole arrays to Si and SiO_2.

15.2 Fabrication

For Ag nanohole arrays attached on Si surface, our approach consists of four steps. First, a drop of polystyrene (PS) sphere with the sphere diameter of 600 nm was deposited on Si substrate. Then it is proceeded to spin coating at a speed of 800 rpm to obtain a self-assembling hexagonal close-packed (hcp) single layer of polystyrene (PS) spheres. Second, the diameter of the PS spheres was reduced to different dimension according to our design by reactive ion etching (RIE) with oxygen. These etched PS spheres were served as masks for creating Ag nanoholes structure. In this step, etching time plays an important role in determining the holes' diameter, thus interhole's distance and the optical properties of the nanostructures. In our work, we use etching time of 80 s to obtain nanosphere diameter of 590 nm, 120 s for nanosphere diameter of 560 nm, and 150 s for nanosphere diameter of 530 nm. Third, 20 nm-thick silver film was deposited through the voids between the closely packed sphere arrays with electron beam evaporator. Finally, polystyrene spheres were removed (lift-off) by acetone and deionized water to form a silver nanohole structure.

For Ag nanohole arrays attached on SiO_2 surface, we fabricated two samples, and there were some extra steps compared with Ag hole arrays attached directly on Si. Before depositing PS, 100 nm SiO_2 was deposited on Si substrate by plasma-enhanced chemical vapor deposition (PECVD) for both samples. Then it followed the same procedure as those on Si. After lifting off the PS spheres, one sample still needs one further step, which is dry etching SiO_2 by RIE with CHF_3 using Ag hole arrays as a mask, leaving a nanohole arrays of Ag on top of SiO_2 hole arrays.

To measure the reflective optical spectra of the fabricated Ag hole arrays, a microspectrometer (CRAIC QDI 2010) was employed. A 75 W xenon white source illuminated normally to the Ag hole arrays. Optical images were captured by a CCD camera. A 36× objective lens was used to focus the measurement aperture onto hole arrays and collect the reflected data.

15.3 Results and Discussion

Figure 15.1 shows the scanning electron microscopy (SEM) images of the surfaces at different stages. Figure 15.1(a) is a uniform monolayer pattern formed by self-assembled nanosphere (with sphere diameter of 600 nm). Figure 15.1(b) is a nanosphere pattern after dry etching. The diameter of the nanosphere is reduced according to our requirement. Figure 15.1(c) is a silver nanohole arrays' pattern after lifting off, which is our final structure.

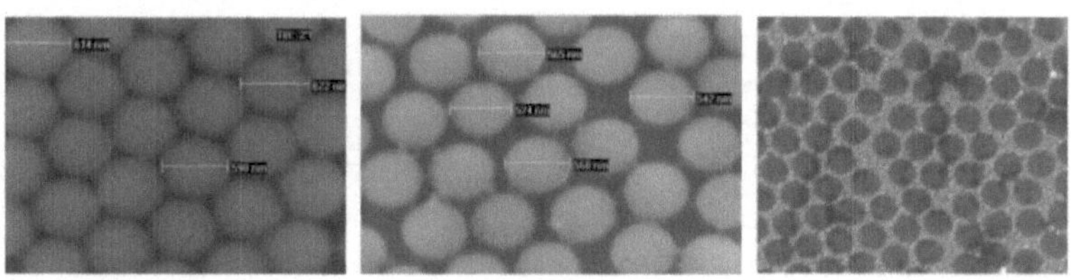

(a) (b) (c)

Figure 15.1 SEM images at different steps during the sample fabrication process. (a) Monolayer nanosphere on substrate. (b) Monolayer nanosphere after dry etching. (c) Ag hole arrays after removing the nanosphere.

Figure 15.2 shows the reflective spectrum of Ag hole arrays with different diameters attached on Si substrate. From the theoretical aspect, plasmonic resonance occurs at the point where absorption reached maximum, and the point where absorption achieved maximum is exactly the same as the point where reflectance reached minimum. So we can identify the plasmonic resonance point from valley of reflective spectrum. It can be observed that plasmonic resonance occurs at different point when varying the diameter of Ag holes. The valley in reflection spectra gradually shifts to longer wavelength when the hole diameter is reduced. For arrays with 590 nm, 560 nm and 530 nm Ag holes, the reflective valleys are located at about 359 nm, 367 nm and 370 nm, respectively, which can be seen clearly from the inset of Figure 15.2.

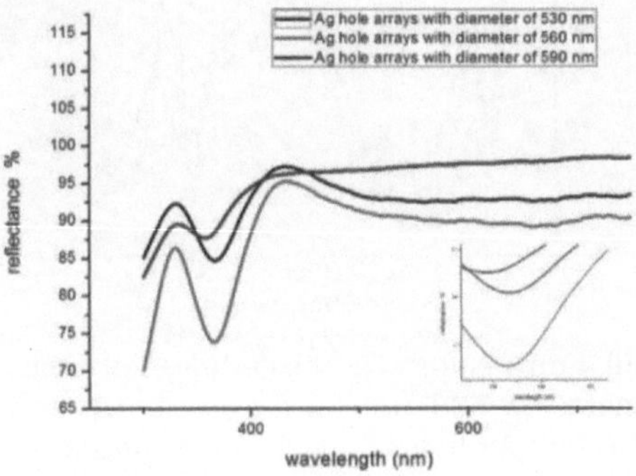

Figure 15.2 Experimental results. The black line represents Ag Holes with diameter of 530 nm based on Si substrate. The red line represents Ag holes with diameter of 560 nm. The blue line represents Ag holes with diameter of 590 nm. Inset shows the enlarged spectrum of plasmonic resonance.

In order to verify the experimental result, simulation was carried out for the same nominal structures. Figure 15.3 shows the simulation results. From Figure 15.3, it is clearly seen that the plasmonic resonance wavelength red-shifts as the hole diameter is decreased, the same trend as our experimental results show. However, the wavelength obtained by the experiment is a bit shorter than those predicted by simulation, as is summarized in Table 15.1. This is due to the fact that the size of the holes cannot be controlled exactly as 530 nm, 560 nm and 590 nm in our experiment, while in simulation exact values are used. Besides, the value of refractive index of Si used in simulation might also be different from the real refractive index of our Si substrate, which will also affect the plasmonic resonance wavelength.

Table 15.1 Comparison between experimental and simulation result

Ag hole diameter (nm)	Plasmonic resonance wavelength (nm)	
	Experimental result	**Simulation result**
590	359	365
560	367	378
530	370	390

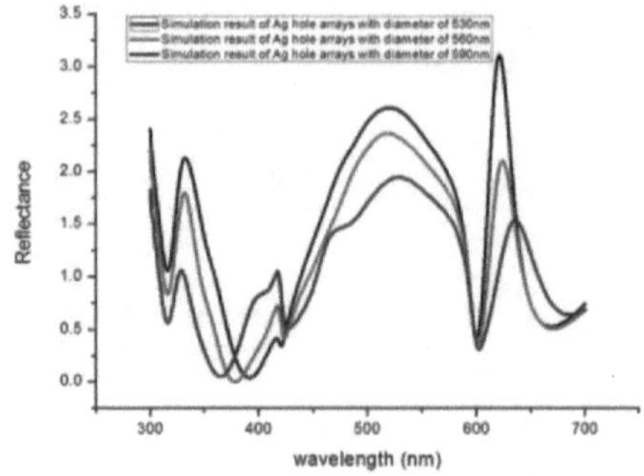

Figure 15.3 Simulation results. The black line represents Ag Holes with diameter of 530 nm based on Si substrate. The red line represents Ag holes with diameter of 560 nm. The blue line represents Ag holes with diameter of 590 nm.

It is also worth pointing out that there are four modes from simulation (Figure 15.3), while only one mode from experimental result is observed (Figure 15.2). This is attributed to the unsuccessful lift-off issue. As it is a challenge to lift off the PS after silver deposition, the surface of sample does not show perfect Ag nanohole arrays. This can be seen from Figure 15.1(c), where some part of the surface is covered by continuous Ag film. The partially continuous Ag film makes the valley of reflective spectrum not low enough, which results in the disappearance of higher order modes. Nevertheless, our experiment shows that the plasmonic resonance wavelength can be tuned by changing the diameter of nanoholes while keeping the array period unchanged.

In order to see how far we can go in tuning the plasmonic resonance wavelength, we also carried out the simulation for a variety of diameters. The software used is Lumerical. The geometry used for simulation was a periodic two-dimensional array of Ag holes with a height of 15 nm. Lateral dimension varying from 590 nm to 300 nm, and a hole-to-hole spacing of 600 nm. The light is incident normally to the Ag holes. Figure 15.4 shows the simulation results of various Ag hole diameter with period of 600 nm and the fitting curve. The black squares are the simulation result for different hole diameters and the red line is the fitting curve for simulation result. From Figure 15.4, we can predict that the plasmonic resonance wavelength can be easily tuned for 100 nm from 360 nm to 460 nm by reducing the hole diameter from 590 nm to 300 nm. Our preliminary experimental result has shown the trend in wavelength tuning. More experiments will be carried out for a wider range of the tuning.

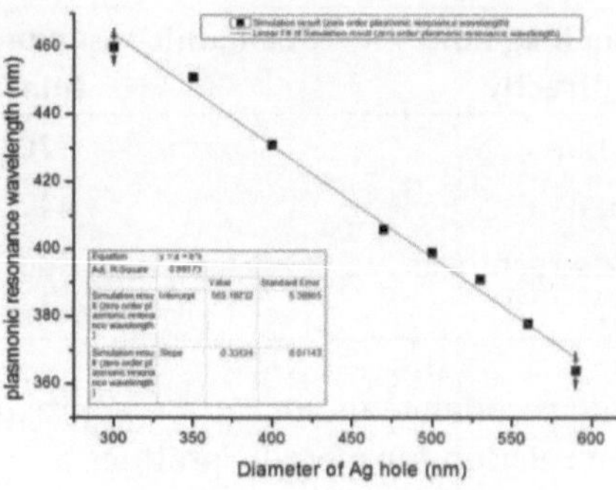

Figure 15.4 Simulation results for various diameters of Ag hole arrays and fitting curve.

We have also studied the effect caused by surrounding medium. Samples with diameter of 530 nm Ag hole arrays attached on SiO_2 flat film and on SiO_2 hole arrays were also measured. Figure 15.5 compares the wavelength tuning caused by the change of surrounding medium.

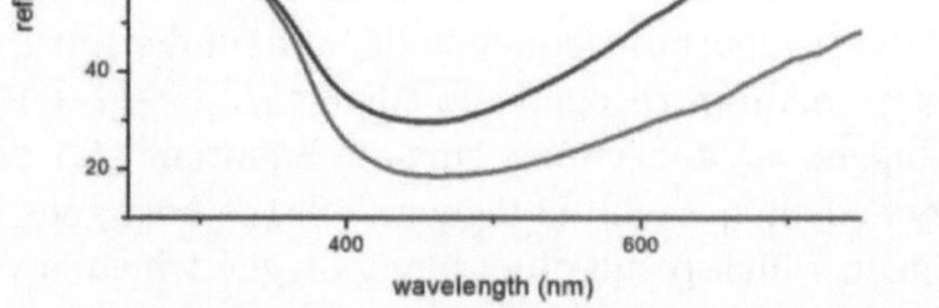

Figure 15.5 The black line represents Ag holes with diameter of 530 nm on Si substrate, whose valley is 370 nm. The red line represents Ag holes with diameter of 530 nm on 80 nm-thick flat SiO_2, whose valley is 447 nm. The blue line represents Ag holes with diameter of 530 nm on 80 nm SiO_2 holes with the same structure as Ag structure based on Si substrate, whose valley is 458 nm.

Table 15.2 Experimental result for Ag nanohole arrays with same diameter (530 nm) and same period (600 nm) but different attached surface

Surface to which Ag hole attached directly	Plasmonic resonance wavelength (nm)
Si flat film	370
SiO_2 flat film	447
SiO_2 hole (same structure as Ag)	450

The plasmonic resonance mode can be approximately given by the SP dispersion relation for metallic gratings at normal incidence [10],

$$\lambda = \frac{d}{\sqrt{i^2 + j^2}} \sqrt{\frac{\varepsilon_s \varepsilon_m}{\varepsilon_s + \varepsilon_m}} \tag{15.1}$$

where ε_s and $\varepsilon_m = \varepsilon_r + \varepsilon_i$ denote the permittivity of substrate media and metal, the integer index i, j corresponds to different sets of valleys, and d is the period of metal nanohole array.

From the theoretical formula, it seems that plasmonic resonance wavelength has no relationship with the hole diameter; only period and permittivity index play determined roles in resonance wavelength. However, experimental results proved that the hole diameter indeed affects the resonance wavelength. We use the Maxwell–Garnett effective-medium theory, equivalent to the dipole–dipole coupling, as a first-order approximation. It predicts that plasmonic resonance occurs at frequency ω, which obeys the condition $\varepsilon_m(\omega) = -\varepsilon_s(2 + P)/(1 - P)$ [11], where P is the volume fraction of the metals, and $\varepsilon_m(\omega)$ and ε_s are the real part of frequency-dependent dielectric constants of metal and medium respectively. Also, $\varepsilon_m/\varepsilon_s = -(2 + P)/(1 - P)$, if P goes bigger, $\varepsilon_m/\varepsilon_s$ becomes smaller. Equation 15.1 shows that if $\varepsilon_m/\varepsilon_s$ gets smaller, λ will be bigger. So if the hole goes smaller, P will be bigger, which results in λ being bigger, which is verified by experimental results correctly.

For different surfaces to which Ag holes attached directly, there exists a big difference in plasmonic resonance point, which can be seen from Figure 15.5. The resonance wavelength of 530 nm Ag holes attached directly on Si is 77 nm shorter than the same Ag structure attached directly on SiO_2 flat surface. The red-shift may be ascribed to the increase of the effective refractive index of surrounding medium when the SiO_2 layer underneath the Ag hole arrays. The structure can also play a role in tuning plasmon resonance wavelength, which can be proved by the experimental result of the same Ag structure attached on flat SiO_2 and SiO_2 hole, respectively. For flat SiO_2 surface on which Ag nanohole array is attached, the plasmonic resonance wavelength red-shifts for 77 nm, while for SiO_2 nanohole arrays beneath the Ag nanohole arrays, the wavelength red-shift for 80 nm, which can be seen from Table 15.2.

15.4 Conclusion

To conclude, we have experimentally demonstrated the tuning of plasmonic resonance wavelength based on Ag hole arrays fabricated by nanosphere lithography technique on Si and SiO_2, respectively. The resonance wavelength red-shifted 11 nm when the Ag hole diameter changes from 590 nm to 530 nm on Si substrate. Our simulation also predicted that the tuning of plasmonic resonance wavelength can be achieved to 100 nm if the diameter of Ag hole arrays were reduced to 300 nm. For the same Ag nanohole arrays attached on Si and SiO_2 surface, respectively, the plasmonic resonance wavelength can be red-shifted for 77 nm. It can further red-shift 3 nm when further etching the planar SiO_2 into nanohole arrays beneath the Ag hole arrays. From the experimental and simulation result, it is clearly proved that the plasmonic resonance wavelength is sensitively dependent on silver nanohole diameter (thus the interhole separation) and its surrounding medium.

References

1. T. W. Ebbesen, H. J. Lezec, H. F. Ghaemi, T. Thio, P. A. Wolff (1998), Extraordinary optical transmission through sub-wavelength hole arrays, *Nature*, 391, 667.
2. J. Boleininger, A. Kurz, V. Reuss, C. Sonnichsen, (2006), Plasmonic materials, *Phys. Chem.*, 8, 324.
3. A. Graff, D. Wagner, H. Ditlbacher, U. Kreibig, (2005), Silver nanowires, *Eur. Phys.* J. D, 34, 263.
4. C. Girard, E. Desjardin, (2006), Near-field optical properties of top-down and bottom-up nanostructures, *J. Opt. A: Pure Appl. Opt.*, 8, S73.

5. A. N. Grigorenko, A. K. Geim, H. F. Gleeson, Y. Zhang, A. A. Firsov, I. Y. Khrushchev, J. Petrovic, (2005). Nanofabricated media with negative permeability at visible frequencies, *Nature*, 438, 335.
6. X. D. Huang, L. R. Bao, X. Cheng, L. J. Guo, S. W. Pang, A. F. Yee, (2002), Reversal imprinting by transferring polymer from mold to substrate,*Vac. Sci. Technol. B*, 2872.
7. T. R. Jensen, M. D. Malinsky, (2000), Nanosphere lithography: tunable localized surface plasmon resonance spectra of silver nanoparticles, *J. Phys. Chem. B*, 104, 10549.
8. A. J. Haes, S. Zou, (2004), Nanoscale optical biosensor: short range distance dependence of the localized surface plasmon resonance of noble metal nanoparticles, *J. Phys. Chem. B*, 108, 6961.
9. M.-D. He, Z.-Q. Gong (2010), Coupling of localized surface plasmon modes in compound structure with metallic nanoparticle and nanohole arrays, *J. Appl. Phys.* 108, 093520.
10. R. Zhou, X. Chen (2008), The surface plasmon resonance of metal-film nanohole arrays, *Solid State Commun.*, 145, 23–28.
11. L. Genzel, T. P. Martin (1973), Infrared absorption by surface phonons and surface plasmons in small crystals, *Surf. Sci.*, 34, 33.

Chapter 16

Revelation of Temperature-Dependent Cathodoluminescence by Employing Nanostructural Pathway in Fe-Incorporated ZnO

Jai Shankar Tawale,[a] Hirotsugu Nishido,[b] Shin Toyoda,[b] Melepurath Deepa,[c] Nupur Bahadur,[a] Mahalingam Maniraj,[d] Sudipta Roy Barman,[d] Bibhash Ranjan Chakraborty,[a] and Avanish Kumar Srivastava[a,*]

[a] *Materials and Chemical Metrology, National Physical Laboratory, Council of Scientific and Industrial Research, New Delhi 110 012, India*
[b] *Department of Applied Physics, Okayama University of Science, 1-1 Ridai, Okayama, 700-0005, Japan*
[c] *Department of Chemistry, Indian Institute of Technology Hyderabad, Hyderabad 502205, India*
[d] *UGC-DAE Consortium for Scientific Research, Khandwa Road, Indore, Madhya Pradesh 452001, India*
*aks@nplindia.ernet.in, avanish.aks555@gmail.com

16.1 Introduction

Recently, fine-scale objects of a variety of materials have been of great interest due to their various tunable microstructures, phase

Advances in Nanodevices and Nanofabrication: Selected Publications from Symposium of Nanodevices and Nanofabrication in ICMAT2011
Edited by Qing Zhang, W. I. Milne, and Jianping Zou

ISBN 978-981-4364-54-6 (Hardcover), 978-981-4364-55-3 (eBook)
www.panstanford.com

transformations and quantum confinements, required for potential use [1–3]. The nanoparticles, wires, rings, combs and tetrapods of semiconducting oxides such as, CdO [4–7], WO_3 [8–10], TiO_2 [11, 12] and ZnO [13–15] are most desired for technological applications. Their novel properties can be tailored significantly by producing them at micro- and nanoscale in different morphologies. Among these, ZnO is a wide-bandgap semiconductor (3.37 eV) with high excitonic binding energy (60 meV). It has a hexagonal (wurtzite type, a = 0.321, c = 0.521 nm) crystal structure with alternating planes of tetrahedrally coordinated O^{2-} and Zn^{2+} ions, stacked along the c-axis, with high density (5606 kg m^{-3}) and melting point (2248 K). It is a promising candidate for short-wavelength optoelectronics, piezoelectric, pyroelectric, solar cells, phototransistors, transparent electrodes and gas sensors. Advancement in the field are directed toward the doping of various transition metal ions such as Mn^{2+} [16, 17], Co^{2+} [18, 19], Ni^{2+} [20], V^{3+} [21, 22] and Fe^{3+} [23–27] into the ZnO lattice to tailor the material microscopically for various optical and electro-magnetic properties.

The aspiration for nanostructure ZnO has been enticing for the luminescent materials since the quantum confinement of excitation may improve the emission efficiency of nanoparticles in comparison to large particles or bulk material [28–34]. Moreover it is worth mentioning that cathodoluminescence (CL) is an extremely sensitive method of measuring the shape, size and dopant-dependent optical response of nanomaterials. In brief, CL is an optical and electrical phenomenon in which a beam of electrons is generated by an electron gun, which then impacts a luminescent material such as a phosphor, causing the material to emit visible light, even by exciting every individual nanoparticle. CL also takes advantage of the morphological information of nanoparticles obtained by using secondary electron (SE) image. At room temperature, ZnO typically exhibits one emission peak in the UV region due to the recombination of free exactions and possibly one or more peaks in the visible spectral range, which are attributed to defect emissions, but the origin of defect emissions is still not fully clear. In addition to experimental studies of the defect emissions, energy levels of different defects and their formation energies have been calculated [35–42]. However, the origins of different defect emissions are still not fully understood, and controversial hypotheses have been proposed to explain the different defect emissions (violet, blue, green, yellow and orange-red). Particular attention is paid here to investigate the cathodoluminescence of undoped and Fe-doped ZnO at room temperature and low temperature. The CL analysis, used together with *in situ* scanning electron microscopy imaging, presents the advantage of high-resolution optical and microstructural characterization. Moreover, lattice scale real and reciprocal space analyses employing high-resolution transmission electron microscopy lead to an authentic corroboration between atomic and electronic structure determination of Fe-influenced ZnO facetted nanostructures.

We have recently reported [43] that the tunable energy bandgaps in the visible region can be significantly obtained via the incorporation of Fe^{3+} in nanostructure ZnO. The dimensional reduction of such materials results in improved non-linearity, controlled by the quantum size effects and other mesoscopic influences. So far, pertaining to fine microstructural features, only limited work has been attempted on the grain growth of Fe-doped ZnO [23–27]. Investigations on different scale microstructures, respective morphologies and crystallographic and lattice scale interpretations of ultra-fine $Zn_{1-x}Fe_xO$ (x = 0.5, 2, 50 mol%) are still warranted. A systematic study on the present work envisages a detailed microscopy even at sub-nanoscale, spectroscopy and optical evaluation of sol-gel derived fascinating objects of Fe-doped ZnO.

16.2 Experimental

Microstructures of Fe (0.5, 2 and 50 mol%)-doped ZnO were synthesized by a simple sol-gel technique using zinc nitrate and ferric chloride as precursors, and subsequently the as-prepared materials were annealed at 800°C for 2 h. The topological features and the composition of Zn, O and Fe were determined by a scanning electron microscope (SEM, model LEO 440) equipped with an energy dispersive X-ray spectrometer (EDS model OXFORD LINK ISIS 300). Microstructural features at high magnifications and selected-area electron diffraction patterns were recorded using high-resolution transmission electron microscopes (HRTEMs, models: JEOL JEM 200CX and Tecnai G^2 F30 STWIN, operated at electron accelerating voltages of 200 and 300 keV, respectively). The x-ray photoelectron spectroscopy (XPS) data were recorded with the specimen mounted on a specially designed sample holder [44] at a base pressure of 3×10^{-10} mbar using an AlKα laboratory x-ray source that was operated at 100 W and a commercial electron energy analyzer with five channeltrons from Specs GmbH. Germany at 50 eV pass energy. The x-ray satellite features due to $AlK\alpha_{3,4}$ have been subtracted using a routine assuming that the satellites produce a replica spectrum of the main line. The inelastic background was subtracted from the raw spectra using the Shirley method.

Absorbance spectra were recorded *in situ* in a glass cell in the 300–1100 nm wavelength range with respect to air in the reference beam in a PerkinElmer Lambda 25 spectrophotometer. A cathodoluminescence detector attached to a scanning electron microscope (SEM, JEOL JSM 7500) was used for recording *in situ* secondary electron (SE) and cathodoluminescence (CL) images and also for acquiring the corresponding energy spectra.

16.3 Results and Discussion

16.3.1 Growth Morphologies and Phase Formations

A set of intriguing growth morphologies of various Fe-doped ZnO samples is depicted in Figure 16.1. In undoped ZnO sample (Figure 16.1a), a prismatic growth of ZnO particles, with a hexagonal basal plane and a conical tip on top has been seen as a normal feature. This prismatic growth is often present with small irregular-shaped fine particles. It is possible to see that the particle growth occurs by layers with very smooth surface and a constant radius of curvature at the growth front. Every layer seems to consist of right hexagonal prismatic crystallites with constant width (Figure 16.1b). It is interesting to note that isolated crystallites on the spherical surface have a predetermined height, possibly governed by the curvature of the underlying layer. On doping (Fe: 0.5, 2, 50 mol%), we observed a systematic variation in powdered microstructures. With 0.5 mol% Fe doping, it was found that the prismatic growth changed to a euhedral shape with a layered structure on their cylindrical surfaces and these layers are almost distributed equidistant starting from basal plane toward the tip of these morphologies (Figure 16.1c). These tips are facetted with hexagonal topography (Figure 16.1c, d). We also see some porosity on the surface of these layered euhedral-type growths. On increasing the dopant concentration to 2 mol%, the porous-structure of euhedral-shaped cylinders with conical tips increases (Figure 16.1e). It is worth mentioning that a fascinating euhedral-shaped morphology in a flower type appearance originating from common centers is an exemplary feature of these samples. In addition to the interesting morphologies, it is evident from these micrographs that the porosity level increases as the dopant concentration increases. However, it was surprising to note that as the dopant concentration increased to 50 mol%, the initial microstructures of prismatic and euhedral morphologies were completely transformed to random shaped particles with some particles of facetted growth (Figure 16.1f). It appeared that as the dopant concentration increased to 50 mol%, probably the porosity in the prismatic and euhedral morphologies was so high that the basic structure (prismatic and euhedral at low Fe content) turned unstable and converted to small almost uniformly distributed random shaped particles. In our previous studies [43], the generation of higher order of porosity and thus the refinement in microstructure with increased dopant concentration have been attributed to the mechanism of grain refinement in several metallurgical systems. In case of excess Fe in ZnO, the solvent matrix (ZnO) pushes the solute (Fe) at the grain boundary and such solute inhibits the grain growth and also leads porosity. It has already been mentioned [45] by XRD results and preliminary TEM observations that the hexagonal-ZnO was present as the dominant phase with the traces of cubic $ZnFe_2O_4$ up to 2 mol% Fe, while at highest 50 mol% Fe doping, $ZnFe_2O_4$ was the observed prominent phase in the microstructure [43]. A detailed electron microscopy exploiting high-resolution imaging was performed on several samples in order to understand nanoscale features, faceting and phase formations in low- and high-content Fe-doped ZnO. As an illustrative example, Figure 16.2 exhibits lattice scale imaging of hexagon-shaped faceted ZnO. A well-defined set of planes oriented along [0001] zone-axis of a hexagonal crystal are seen (Figure 16.2). We do not see any distortion in lattice structure. Corresponding relatively low-magnification micrograph clearly reveals a small facetted particle with six equally distant (~ 20 nm) and equally angular faces with sharp edges and vertices (inset 'a' in Figure 16.2). Inset 'b' in Figure 16.2 shows a corresponding selected area electron diffraction pattern (SADP) along [0001] zone axis of hexagonal-ZnO. A set of important planes (hkil: $\bar{1}100$, $01\bar{1}0$, $10\bar{1}0$) are marked on SADP. Another high-resolution image (Figure 16.3) recorded on sample 50 mol% Fe-doped ZnO revealed a region consisting of nanocrystallites of cubic $ZnFe_2O_4$ grown along different orientations with a majority along 311 (d = 0.24 nm) along with other directions of 220 (d = 0.29 nm) and 400 (d = 0.15 nm). A corresponding SADP of this region shows Debye rings exhibiting the planes: 220, 311, 400, 511 and 440 (inset in Figure 16.3). It is important to mention that in between low-Fe-content (0.5, 2 mol%) ZnO exhibited various nanoscale features. As an illustrative example, Figure 16.4a depicts a fine grain of about 50 nm of cubic $ZnFe_2O_4$ stacked along 311 planes with interplanar spacing of 0.24 nm. Figure 16.4b shows a coexistence of hexagonal ZnO and cubic $ZnFe_2O_4$ phases and elucidates the planes along 0002 of ZnO and 311 of $ZnFe_2O_4$. The interface between ZnO and $ZnFe_2O_4$ appears clean except that a nominal diffuse contrast and stacking coherency due to the difference in interplanar spacings of ZnO (d = 0.25 nm) and $ZnFe_2O_4$ (d = 0.24 nm) by a lattice mismatch of 4%. At the interface, an angular separation of about 139°, between the two phases, was also noticed.

Figure 16.1 SEM micrographs recorded from the samples: (a,b) bare ZnO (c,d) 0.5% Fe doped-ZnO, (e) 2% Fe-doped ZnO and (f) 50% Fe-doped ZnO.

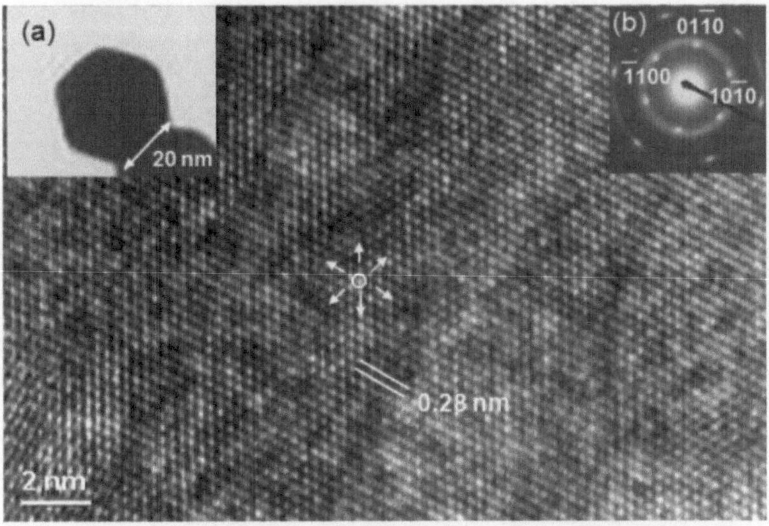

Figure 16.2 High-resolution TEM micrograph showing the lattice scale image of bare-ZnO along the *c*-axis of hexagonal crystal structure. Insets; (a) ZnO crystal along c-axis, (b) corresponding electron diffraction pattern along [0001] zone axis.

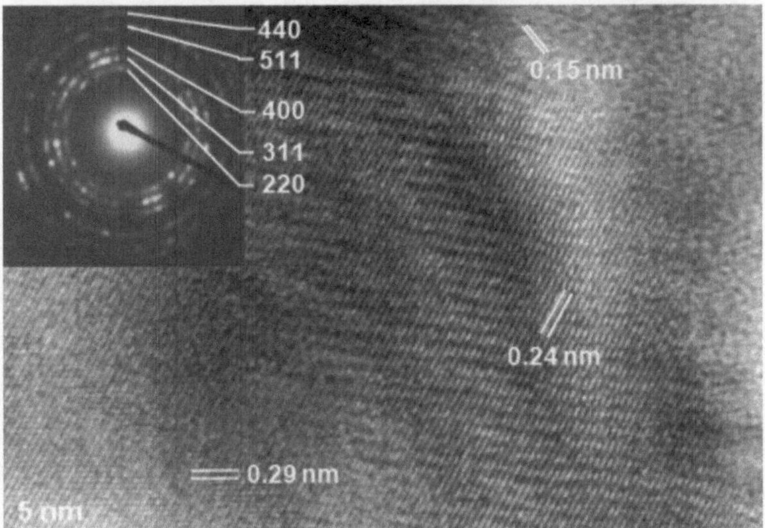

Figure 16.3 High-resolution TEM micrograph showing the lattice scale image of different nanoscale crystallites of cubic $ZnFe_2O_4$. Corresponding electron diffraction pattern showing Debye rings of different planes of cubic phase.

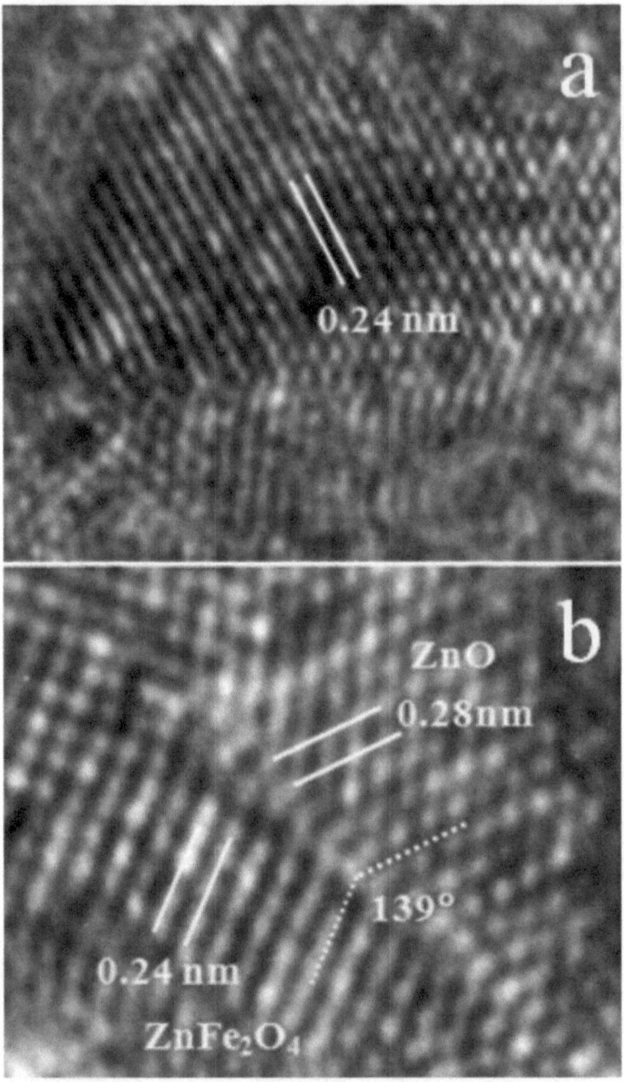

Figure 16.4 High-resolution TEM micrograph showing the lattice scale images of (a) cubic $ZnFe_2O_4$ and (b) interface between cubic $ZnFe_2O_4$ and hexagonal ZnO.

The high-resolution image (Figure 16.2) of ZnO was recorded with the incident electron beam exactly parallel to the six-fold symmetry zone axis [0001] of hexagonal phase. The image shows a distribution of bright dots lying along the six-fold directions with an angle of 2Π/6 rad as indicated by arrows, with a perfect translational symmetry (Figure 16.2). As we continue on adding Fe into the lattice of (Zn,Fe)O, the formation of $ZnFe_2O_4$ may be understood as the precipitation from supersaturated solid solutions. The present observations show that on adding high Fe concentration to bare hexagonal ZnO and/or low-Fe-content (0.5, 2 mol%) ZnO, i.e., (Zn,Fe)O, also referred as Zn(Fe)O (this terminology symbolizes that the Fe is introduced in basic ZnO lattice without changing the original hexagonal structure of bare ZnO; however, the distortions in basic hexagonal are always possible on introducing Fe) leads to a cubic $ZnFe_2O_4$ as a dominant phase in the microstructure. The crystals with cubic structure grow into the shapes of facetted fine particles (cuboids) of dimensions about 30 to 50 nm. Figure 16.5 elaborates the transformation mechanism and illustrates the prismatic growth in bare ZnO particles with hexagonal basal plane and conical tip (Figure 16.5a). Low Fe doping of 0.5 mol% led to a systematic variation in microstructure from individual entity of prismatic growth to euhedral shape morphology with well define and equidistant layers with a flowery appearance having a common base (Figure 16.5b). As shown in Figure 16.5c, 2 mol% Fe doping led to disintegration of the flower shape morphology, which was driven by the formation of dislocation networks, that can be well correlated with the development of ionized vacancies (O^- and Zn_i), which resulted in the origin of the green and blue emission, respectively, at 0.5 and 2 mol% Fe doping [43]. Furthermore, this disintegration resulted in the formation of structural defects/porosity at the surface of these particles, which led to instability in the micro-structure. The trend continued until the segregation of excessive $ZnFe_2O_4$ phase from the host matrix of ZnO at 50 mol%, resulting in the complete transformation of the fragmented particles to random shape morphologies of different polyhedral (Figure 16.5d).

It is understood that the hexagonal close packed (hcp) structure has a primitive cell of hexagonal lattice with a basis of two atoms but the fcc (cubic) primitive cell has a basis of one atom. The ratio c/a for hcp of spheres has the value 1.633. The number of nearest neighbor atoms in both hcp and fcc structure is same, i.e., 12, and if the binding energy depends on the nearest neighbor bond per atom, then there would be no difference in energy between hcp and fcc structures. Crystallographically, a transformation between hexagonal (ZnO) to cubic ($ZnFe_2O_4$) is possible because the planes of hexagonal and cubic along 110 direction can be twin related. The hexagon along the basal axis is possible to evolve by a set of 110-cubic planes [46]. This twinning mechanism satisfies a probable epitaxial condition to grow cubic phase on hexagonal by attaining more thermodynamically stable system, as the cubic phase is crystallographically most symmetric structure. It is also reasonable to expect a certain relaxation of the structure as a consequence of the fruition of cubic phase over hexagonal. The epitaxial growth is evidenced from a high-resolution micrograph (Figure 16.4b) that exhibits a good matching between crystal planes of ZnO and $ZnFe_2O_4$ on either side of the interface.

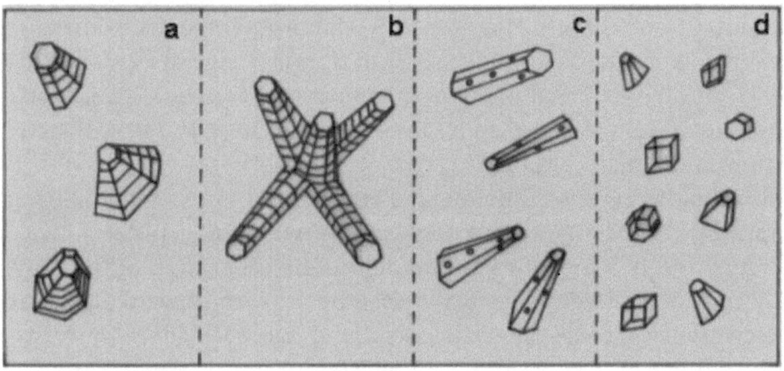

Figure 16.5 Schematic showing the different steps of nanostructural transformations on varying Fe content: (a) 0, (b) 0.5, (c) 2 and (d) 50 mol% in ZnO matrix.

The angular separation (139°) observed in this micrograph clearly reveals that a small mismatch of 4% between the interplanar spacing of ZnO and $ZnFe_2O_4$ causes the deviation from ideal 180° between the crystal planes of two phases at a common boundary.

16.3.2 Spectroscopy and Emission Bands

X-ray photoelectron spectroscopy (XPS) analysis was carried out on doped samples to analyze the chemical and electronic state of Fe under different concentration in ZnO matrix. The XPS survey spectra and the Fe 2p core level spectra for the Fe-doped ZnO samples are displayed as shown in Figures 16.6A and 16.6B, respectively. The C1s correction has been applied to the core level spectra shown in

Figure 16.6. While the wide scans for 0.5 and 2 mol% Fe-doped ZnO (Figure 16.6A(a,b)) are almost similar, the spectrum obtained for the sample with a substantial (50 mol%) doping level of iron is quite different (Figure 16.6A(c)). For the latter sample, the characteristic signal of Zn $2p_{3/2}$ at 1022 eV binding energy (BE) is significantly suppressed and the Fe $2p_{3/2}$ signal at 710 eV is intense and prominent. On the other hand, the reverse is observed for the samples with lower doping levels of 0.5 and 2 mol% of Fe. This implies that the intensity of the Fe core level is substantial only for the sample with a high Fe content. This is also confirmed from the core level spectra shown in Figure 16.6B, wherein the signals due to Fe 2p core levels are exceedingly weak for the samples containing 0.5 and 2 mol% of Fe (Figure 16.6B(a′,b′)). The BE of the Fe $2p_{3/2}$ and Fe $2p_{1/2}$ spin-orbit components of the sample with 50 mol% Fe doping level are 711.6 and 724.6 eV, respectively. The observed Fe $2p_{3/2}$ peak position (711.6 eV) agrees with the BE of Fe 2p in Fe^{3+} state observed for Fe_2O_3 [47]. Its position is significantly different from FeO (Fe^{2+} state) that is expected at 709.6 eV BE, i.e., at 2 eV lower BE with respect to the Fe^{3+} state. The evidence of the Fe^{2+} state has been observed in an earlier study of Fe-doped ZnO as a shoulder in the lower BE side of the main peak [48]. However, here we do not observe such a shoulder, which indicates the predominance of the Fe^{3+} state. This is further supported by the position of the satellite peaks that are at about 8 eV toward the higher BE side of the Fe $2p_{3/2}$ and $2p_{1/2}$ main peaks (Gaussian components 5 and 10 in Figure 16.6B(c′)). Occurrences of satellites as well as multiplet component peaks are well known in the XPS spectra of transition metal oxides [49]. The satellite peaks in this case originate from the $2p^5 3d^5$ final state of the Fe^{3+} and are in good agreement with literature [50].

The multiplet component peaks in Fe 2p core-level spectrum for Fe^{3+} valence state have been determined from x-ray absorption and high-resolution photoemission study [50]. Although these components cannot be identified in the present spectrum (Figure 16.6B(c′)) due to limitation of the resolution, we have fixed the BE position of the 8 multiplet components from Ref.[50] and represented them by Gaussians (1–4, 6–9) to perform least square fitting of the Fe 2p spectrum using the Levenberg–Marquardt algorithm, as in our previous work. [44]. The quality of fit is reasonably good (blue dashed line), but there is a small mismatch in the lower BE side,

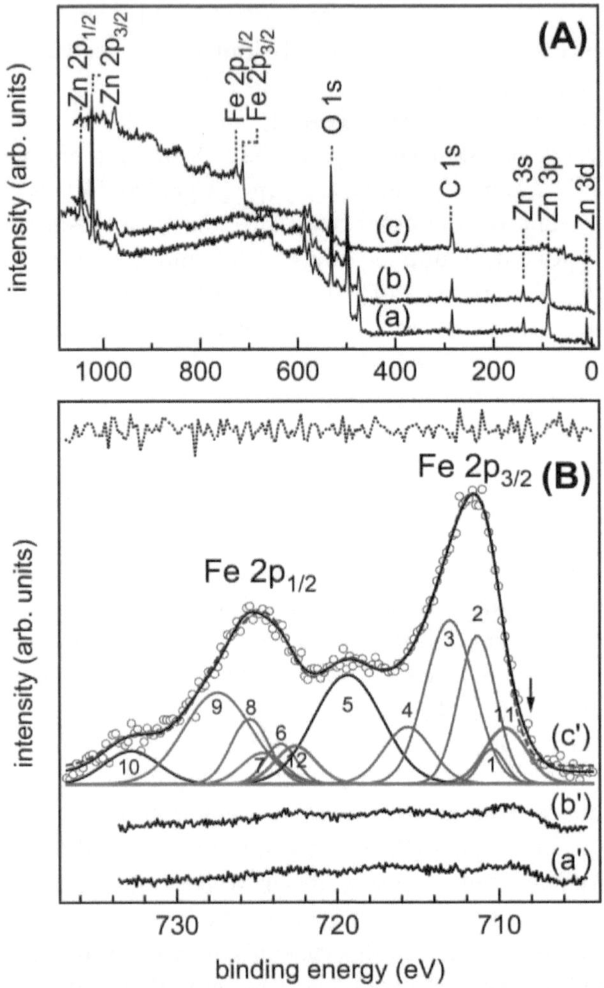

Figure 16.6 (A) XPS survey spectra of ZnO doped with (a) 0.5, (b) 2 and (c) 50 mol% of Fe and (B) Fe 2p core level spectra for doping levels of (a′) 0.5, (b′) 2 and (c′) 50 mol% of Fe. The Fe 2p spectrum for doping level of 50′mol% of Fe has been fitted with Gaussians (red line) 1–4 and 6–9 that correspond to multiplet component peaks of the Fe3+ valence state, while Gaussians 5 and 10 are the satellite features. Gaussians 11 and 12 (green bold line) represent the Fe2+ valence state. The fitted curve (dashed blue line) through the data points (open circles) is obtained using Gaussians 1–10, while the fitted curve (black solid line) and residual (dotted line) are obtained using Gaussians 1–12 (cf. see text).

as shown by an arrow. To examine whether this could indicate the presence of a small amount of Fe^{2+}, we have used two more Gaussians 11–12 (green bold lines) at 709.6 and 722.6 eV to represent the Fe^{2+} state spin-orbit components. The fitting has been performed again without changing any of the peak positions but allowing their intensities to vary. The quality of the fit improves (black solid line) in the lower BE region in particular. All the Gaussian components of the fit are shown in Figure 16.6B. As expected, the ratio of the Gaussians 12 and 11 representing Fe^{2+} $2p_{1/2}$ and $2p_{3/2}$ spin orbit states, respectively, is 0.67, which is very similar to that of the Fe^{3+} spin orbit components (0.66). Another important point that justifies the existence of small Fe^{2+} component is that only when Gaussians 11 and 12 are considered, the relative intensities of the Fe^{3+} multiplet peaks (Gaussians 1–4 and 6–zx agrees with literature [50]. We can estimate the ratio of Fe^{2+}/Fe^{3+} valence states by dividing the area of Gaussian 11 (Fe^{2+}) by the added area of Gaussians 1–4 (Fe^{3+}) to be about 0.14, i.e., about 12% of Fe is in 2^+ state.

Thus, the XPS results are in agreement with the finding that $ZnFe_2O_4$ (where Fe is in 3^+ state) is a dominant phase in the microstructure for the sample with 50 mol% Fe doping. Other reasons for the predominance of the Fe^{3+} state in the XPS spectrum could be related by the presence of Zn vacancies or excess oxygen at the surface [48].

Figure 16.7 shows the nature of the cathodoluminescence (CL) and corresponding secondary electron (SE) images of Fe-doped ZnO material under SEM. The specimens were illuminated at the same area to obtain the images from SE and CL modes. The bare ZnO exhibits the luminescence at grain boundaries and edges of the hexagonal plane of ZnO due to intrinsic impurities and defects in the sample. Fe-doped ZnO CL images show systematic variation in the microstructure and impact of doping on the luminescence. In case of low Fe concentration (0.5 and 2 mol%), CL emission increases on surface, grains boundaries and distributed uniformly across sample. The physical resolutions of the images are same, but CL resolves the grain boundary, which is not possible with SEM. This helps resolve clearly hexagonal faces of the particles (shown in 2 mol% CL image). On further increase in Fe concentration up to 50 mol%, the shape of the particles change to nanocuboids and CL emission reduces across the sample with porosity in between the grains.

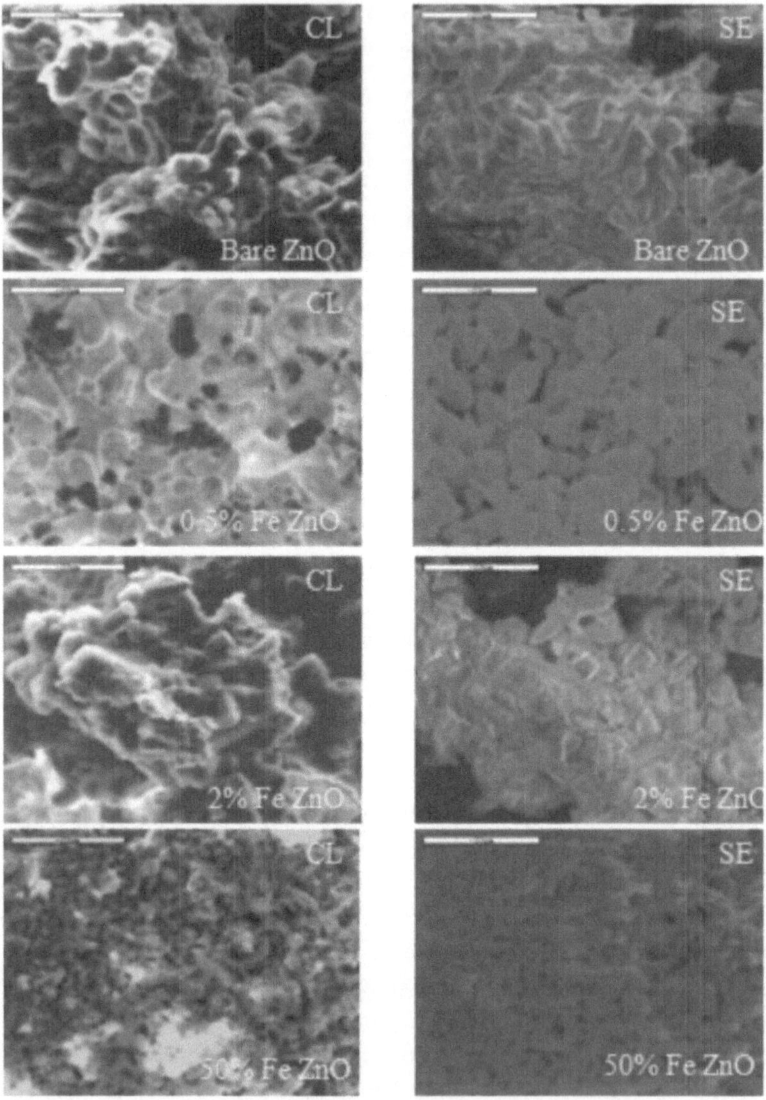

Figure 16.7 CL and SEM images of undoped and 0.5, 2 and 50 mol% Fe-doped ZnO (a) CL image (b) SEM image.

The room temperature CL spectra of Fe (0, 0.5, 2, and 50) mol% doped ZnO material are depicted in Figure 16.8. The ZnO shows CL emission centered at around 392 nm (~ 3.13 eV) corresponds to near band edge emission (NBE) pertaining to excitonic recombination of electrons and photogenerated holes [49] and broad orange emission around 635 nm (~ 1.94 eV). In Fe-doped ZnO samples, the intensity of NBE emission reduces with respect to increased doping. In 0.5 mol% Fe-doped ZnO, CL spectra were taken at bright (2a) and dark regions (2b). The bright region shows similar broad orange emission as in case of (2a). With 2 mol% strong CL orange emission at 635 nm (~1.94 eV) was observed, whereas at 50 mol% Fe doping no CL emission across the sample at room temperature.

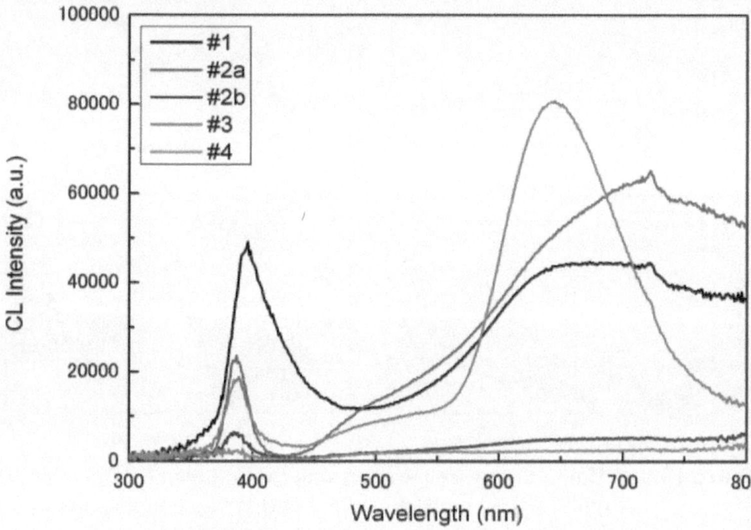

Figure 16.8 CL spectra of (#1) undoped ZnO, (#2) 0.5% Fe-doped ZnO, (#3) 2% Fe-doped ZnO,(#4) 50% Fe-doped ZnO at room temperature.

CL response of Fe (0, 0.5, 2, and 50) mol% doped ZnO structures with respect to change in low temperature from (–50, –100 and –190)°C are shown in Figure 16.9. The bare ZnO sample shows CL emission peak centered on 392 nm (~3.13 eV) (same as at room temperature) corresponds to NBE peak and it reduces with decrease in temperature from (–50 to –190)°C (Figure 16.9a). With Fe doping the NBE intensity peak decreases with increase in Fe concentration and multiple peaks are observed in the visible region. In Figure 16.9b, Fe 0.5 mole% shows CL emission centered around 535 nm (~2.29 eV) and 632 nm (~1.94 eV) correspond to green and orange emission, respectively. On 2 mol% Fe doping, Figure 16.9c shows only green emission at 535 nm (~2.29 eV). Intensity of the CL emission decreases from (–190 to –50)°C up to 2 mol% Fe. At maximum 50 mol% Fe doping Figure 16.9d shows green emission with intensity trend exactly reverse of previous and it decrease with change in temperature from (–50 to –190)°C like undoped ZnO.

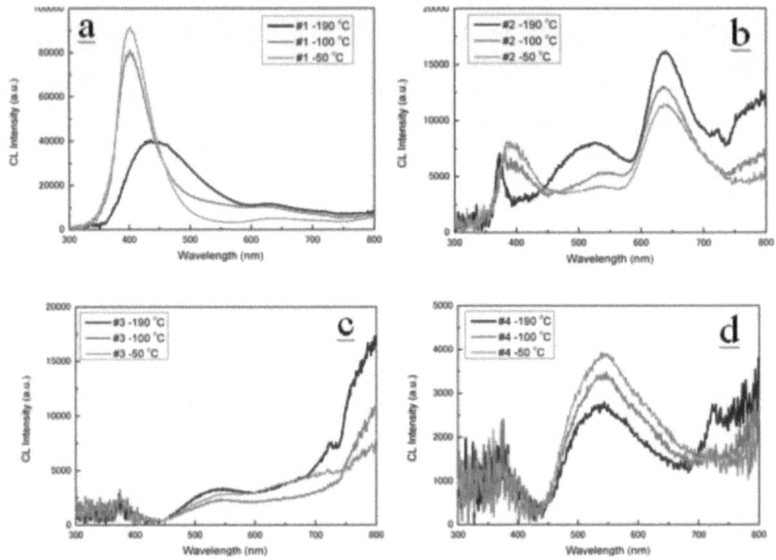

Figure 16.9 Cathodoluminescence spectra of (a) undoped (b) 0.5 mol% (c) 2 mol% (d) 50 mol% Fe-doped ZnO at low temperatures.

CL is the complex function of concentration, lattice structure and superimposed strain on the structure of the material due to temperature and doping concentration. The CL micrographs of Fe-doped ZnO reveals luminescence is located at definite sites giving grain like a structure inherent to the sample surface morphology. The luminescence centers result from impurities in the structure and the impurities are Fe^{3+} ion in $ZnFe_2O_4$ lattice structures. Fe incorporates the additional luminescence center in ZnO lattice structure due to the inclusion of extra energy states between conduction and valence bands as confirmed by photoluminescence too. Fe^{3+} ions in ZnO structure form $ZnFe_2O_4$ phase, and the geometry of the $ZnFe_2O_4$ phase formation has a significant effect on the CL response. The distribution of CL in Fe-doped ZnO provides fundamental insights in the process, such as crystal growth, replacement and deformation.

Low-temperature CL spectroscopy illuminates orange and green emission bands in the visible region. At higher Fe content, the $ZnFe_2O_4$ phase formation may lead to excess singly ionized oxygen vacancies responsible for green emission and double-ionized oxygen vacancies are replaced by Fe^{2+} ions result in to a no longer orange emission at room as well as low temperature. We have demonstrated that Fe-incorporated ZnO materials are useful for defect emission study and application in optoelectronic devices.

16.4 Conclusions

Sol-gel-derived ZnO nanoparticles in bare and in Fe-incorporated conditions have been examined in depth by high-resolution electron microscopy, and a microstructural pathway has been established with the content of Fe in the basic ZnO-matrix. The microscopy and spectroscopy observations sufficed the explanation of alterations in cathodoluminescence at different temperatures with the Fe incorporation in ZnO.

Acknowledgments

The Indo-Japanese collaborative project (DST ref. no. DST/INT/JAP/P-51/08 and JSPS ref. no. 08039221-000441) is gratefully acknowledged. The authors (NB, AKS) thank DST for financial support (Grant no.SR/FTP/CS-57/2006). Dr. B. R. Chakraborty is thankful to CSIR for financial assistance under emeritus scientist scheme.

References

1. H. Gleiter, *Acta Mater.*, 48 (2000) 1.
2. H.S. Nalwa, *Handbook of Nanostructured Materials and Nanotechnology*, Academic Press Publishing, Tokyo, 2000.
3. F. Carlier, S. Benrezzak, Ph. Cahuzac, N. Kébaïli, A. Masson, A.K. Srivastava, C. Colliex, C. Bréchignac, *Nano Lett.*, 6 (2006) 1875.
4. A.K. Srivastava, S. Pandey, K. Lal, K.N. Sood, R. Kishore, *Proceedings of the Conference on Electron Microscopy and XXVIII Annual Meeting of EMSI*, Thiruvananthapuram, India, 2006, p.66.
5. D. Ma, Z. Ye, L. Wang, J. Huang, B. Zhao, *Mater. Lett.*, 58 (2003) 128.

6. Z.R. Dai, Z.W. Pan, Z.L. Wang, *Adv. Funct. Mater.*, 13 (2003) 9.
7. D.M.C. Galcia, R.C. Pérez, O.J. Sandoval, S.J. Sandoval, G.T. Delgado, C.I.Z. Romero, *Thin Solid Films*, 371 (2000) 105.
8. A.K. Srivastava, S.A. Agnihotry, M. Deepa, *Thin Solid Films*, 515 (2006) 1419.
9. M. Deepa, A.K. Srivastava, K.N. Sood, S.A. Agnihotry, *Nanotechnology*, 17 (2006) 2625.
10. M. Deepa, A.K. Srivastava, S.A. Agnihotry, *Acta Mater.*, 54 (2006) 4583.
11. J.M. Wu, H.C. Shih, W.T. Wu, *Nanotechnology*, 17 (2006) 105.
12. S. Bhandari, M. Deepa, A.K. Srivastava, H.-J. Kleebe, S.T. Lakshmikumar, A. Ramakant, *J. Nanosci. Nanotechnol.*, (2006), communicated.
13. Z.L. Wang, *J. Phy: Condens. Matter*, R829 (2004) 16.
14. A.K. Srivastava, K.N. Sood, K. Lal, R. Kishore, Patent filed ref. no. 0773DEL2005 dated 31 March, 2005.
15. A.K. Srivastava, N. Gupta, K. Lal, K.N. Sood, R. Kishore, *J. Nanosci. Nanotechnol.*, 7 (2007) 1.
16. G-R. Li, D-L. Qu, W-X. Zhao, Y-X. Tong, *Electro. Commun.*, 9 (2007) 1661.
17. S. Deka, P.A. Joy, *Solid State Commun.*, 142 (2007) 190.
18. K.J. Kim, Y.R. Park, *Appl. Phys. Lett.*, 81 (2002) 1420.
19. X-C Liu, E-W Shi, Z-Z Chen, H-W Zhang, L-X Song, H. Wang, S-D Yao, *J. Crystal Growth*, 296 (2006) 135.
20. E. Liu, P. Xiao, J.S. Chen, B.C. Lim, L. Li, *Curr. Appl. Phys.* In Press, Corrected Proof, Available online 25 October 2007.
21. H. Saeki, H. Tabata, T. Kawai, *Solid State Commun.*, 120 (2001) 439.
22. M.E. Koleva, P.A. Atanasov, N.N. Nedialkov, H. Fukuoka, M. Obara, *Appl. Surf. Sci.*, 254 (2007) 1228.
23. N.R.S. Farley, C.R. Staddon, L. Zhao, K.W. Edmonds, B.L. Gallagher, D.H. Gregory, *J. Mater Chem.*, 14 (2004) 1087.
24. S. Baek, J. Song, S. Lim, *Physica B*, 399 (2007) 101.
25. A. Debernardi, M. Fanciulli, *Physica B*, 401–402 (2007) 451.
26. A.J. Chen, X.M. Wu, Z.D. Sha, L.J. Zhug, Y.D. Meng, *J. Phys. D: Appl. Phys.*, 39 (2006) 4762.
27. Y.Q. Wang, S.L. Yuan, L. Liu, P. Li, X.X. Lan, Z.M. Tian, J.H. He, S.Y. Yin, *Magn. Magn. Mater.* In press, doi:10.1016/j.jmmm.2007.10.007 (2007).

28. A.B. Djurišić, Y.H. Leung, *Small*, 2 (2006) 944.

29. L.E. Greene, M. Law, J. Goldberger, F. Kim, J.C. Johnson, Y. Zhang, R.J. Saykall, P. Yang, *Angew. Chem. Int. Edn.*, 42 (2003) 3031.

30. W.M. Kwok, Y.H. Leung, A.B. Djurišić, W.K. Chan, D.L. Phillips, *Appl. Phys. Lett.*, 87 (2005) 093108.

31. A.B. Djurišić, Y.H. Leung, K.H. Tam, L. Ding, W.K. Ge, H.Y. Chen, S. Gwo, *Appl. Phys. Lett.*, 88 (2006) 103107.

32. R.B.M. Cross, M.M. De Souza, E.M.S. Narayanan, *Nanotechnology*, 16 (2005) 2188.

33. M.H. Huang, S. Mao, H. Feick, H. Yan, Y. Wu, H. Kind, E. Weber, R. Russo, P. Yang, *Science*, 292 (2001) 1897.

34. B.P. Zhang, N.T. Binh, K. Wakatsuki, Y. Segawa, Y. Kashiwaba, K. Haga *Nanotechnology*, 15 (2004) S382.

35. B. Lin, Z. Fu, Y. Jia, *Appl. Phys. Lett.*, 79 (2001) 943.

36. P.S. Xu, Y.M. Sun, C.S. Shi, F.Q. Xu, H.B. *Pan Nucl. Instrum. Methods*, B 199 (2003) 286.

37. S.A.M. Lima, F.A. Sigoli, M. Jafelicci Jr, M.R. Davolos, *Int. J. Inorg. Mater.*, 3 (2001) 749.

38. A. Janotti, C.G. Van de Walle, *J. Cryst. Growth*, 287 (2006) 58.

39. J.L. Zhao, W. Zhang, X.M. Li, J.W. Feng, X. Shi, *J. Phys.: Condens. Matter*, 18 (2006) 1495.

40. A.F. Kohan, G. Ceder, D. Morgan, C.G. van de Walle, *Phys. Rev. B*, 61 (2000) 15109.

41. S.B. Hang, S.H. Wei, A. D Zunger, *Phys. Rev. B*, 63 (2001) 075205.

42. F. Oba, S.R. Nishitani, S. Isotani, H. Adachi, I.J. Tanaka., *Appl. Phys.*, 90 (2001) 824.

43. A.K. Srivastava, M. Deepa, N. Bahadur, M.S. Goyat, *Mater. Chem. Phys.*, 114 (2009) 194.

44. R.S. Dhaka, A.K. Shukla, M. Maniraj, S.W. D'Souza, J. Nayak, S.R. Barman, *Rev. Sci. Instrum.*, 81 (2010) 043907.

45. A.K. Srivastava, K. Ninagawa, S. Toyoda, B.R. Chakraborty, *Opt. Mater.*, 32 (2009) 410.

46. S. Zhou, K. Potzger, H. Reuther, G. Talut, F. Eichhorn, J. Borany, W. Skorupa, M. Helm, J. Fassbender, *J. Phys. D: Appl. Phys.*, 40 (2007) 964–969.

47. T. Fujii, *et al.*, *Phys. Rev. B*, **59**, 3195 (1999).

48. T. Kataoka, *et al.*, *J. Appl. Phys.*, **107**, 033718 (2010).

49. Z.Y. Wu, M. Benfatto, M. Pedio, R. Cimino, S. Mobilio, S.R. Barman, K. Maiti, D.D. Sarma, *Phys. Rev. B*, 56, 2228 (1997).

50. T. Droubay, S.A. Chambers, *Phys. Rev. B*, 64, 205414 (2001).

Chapter 17

Raman and Photoemission Spectroscopy Studies on Thermal Stability of Quantum Dot Consisting of Si Clad and Ge Core

Yudi Darma
Physics Department, Institut Teknologi Bandung , Ganesa 10 Bandung, 40132, Indonesia
yudi@fi.itb.ac.id

The compositional mixing and crystallinity of Si dots with Ge core (~20 nm in average dot diameter), which were prepared on thermally grown SiO_2 by highly selective low-pressure chemical vapor deposition (LPCVD), have been studied as a function of annealing temperatures in the range of 540–1000°C. XPS analysis of annealed single layer structures confirms the incorporation of Ge atoms into Si clad from Ge core and the formation of GeO_x at the Si clad surface. Raman scattering spectra of multiply stacked structures consisting of dots with Ge core and 2 nm-thick SiO_2 interlayers indicate the compositional mixing occurs partly at Si/Ge interface during the sample preparation, where the sample was heated up to 600°C. Also, 800°C annealing promotes Si–Ge alloying between Si clad and Ge core and degrades the crystallinity. By 950°C annealing,

Advances in Nanodevices and Nanofabrication: Selected Publications from Symposium of Nanodevices and Nanofabrication in ICMAT2011
Edited by Qing Zhang, W. I. Milne, and Jianping Zou

ISBN 978-981-4364-54-6 (Hardcover), 978-981-4364-55-3 (eBook)
www.panstanford.com

SiO_2 reduction with diffused Ge atoms to form volatile GeO_x units becomes significant, resulting in the generation of Si–Si bonds.

17.1 Introduction

Nanometer-size heterostructure dots with a Si/Ge stacking structure have attracted extensive interest because of their unique properties associated with carrier confinement [1]. The thermal stability of SiGe/Si and Ge/Si interfaces has so far been studied in $Si_{1-x}Ge_x$/Si superlattices [2], SiGe nano-islands embedded in Si [3] and Ge nano-islands on Si [4]. It has been reported that the interdiffusion coefficient of the nano-islands is much higher than that of layered structures and also atomic interdiffusion can decrease the strain energy.

For precise control of the three-dimensional size effect in the dots, annealing temperature dependence of the intermixing and crystallinity has yet to be studied in detail. We have studied self-assembling of nanometer Si dots on SiO_2 by controlling the early stages of LPCVD of monosilane and also demonstrated the spatial arrangement and positioning of Si dots on SiO_2 [5]. We have extended our research to the formation of Si dots with Ge core by alternately controlling the selective growth conditions in LPCVD using monosilane (SiH_4) and 5% Germane (GeH_4) diluted by He [6,7].

In this chapter, we focus the characterization of the compositional intermixing in the Si dots with Ge core and the crystallinity as a function of annealing temperatures using Raman and X-ray photoemission spectroscopy.

17.2 Experimental Detail

Hemispherical single-crystalline Si dots were first grown on thermally grown SiO_2 on Si(100) at 560°C from the thermal decomposition of pure SiH_4 by maintained the gas pressure at 0.5 Torr. Ge deposition was performed selectively on pregrown Si dots at 400°C for 3 min using 5% GeH_4 diluted with He and subsequently followed by the Si selectively deposition on the dots at 540°C under a SiH_4 pressure of 0.02 Torr. To minimize the compositional mixing during this preparation process, after each deposition step, CVD chamber was completely purging with dry N_2 and evacuating down to ~10^{-5} Torr. For preparing multiply stacked dots structures, surface oxidation of the Si cap was carried out at 600°C to form ~2 nm-thick SiO_2. Isochronal annealing for 5 min each was performed in N_2 ambient in the temperature range of 800–1000°C and after each annealing step, Raman scattering measurements were carried out in Ar ambient at a right-angle scattering geometry.

17.3 Result and Discussion

Cross-sectional TEM images show spherical nanometer dots made of Si clad and ellipsoidal Ge core (Figure 17.1), in contrast to hemispherical pure-Si dots pregrown on SiO_2. This implies that the strained energy is larger than the bonding energy at Si/SiO_2 interface generated at the Si/Ge interface. Notice that the crystallographic orientation of the Si cap is different from that of pregrown Si dot presumably due to the structural strain at the Si/Ge interface. The orientation of Ge core is not clear because of poor contrast.

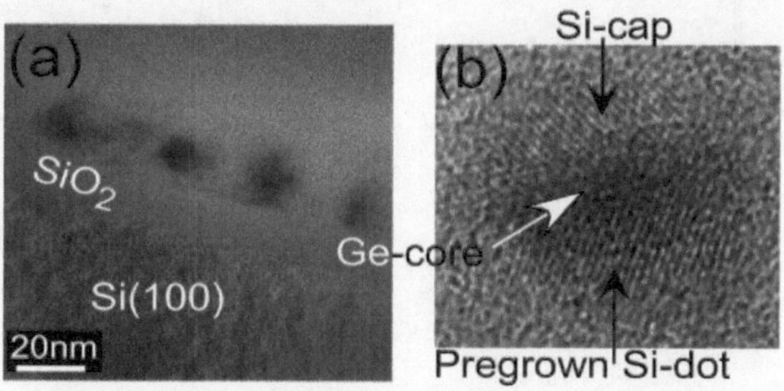

Figure 17.1 Cross-sectional HR-TEM images of Si dots with Ge core, (a) an as-grown sample with isolated dots (b) an enlarged image of single dot.

Raman scattering spectra for as-prepared samples (Figure 17.2a) imply that the compositional mixing partly occurs at Si/Ge interfaces during LPCVD as verified from relative intensity of Si–Ge bonds with respect to Si–Si and Ge–Ge photon intensity. The ~490 cm^{-1} peak seen in the Raman spectrum of the sample annealed at 850°C indicates the intermixing between the Si clad and the Ge core. With 950°C annealing, the compositional intermixing seems to spread all over the Si clad. A significant increase in the Si–Si TO phonon intensity with 1000°C annealing suggests that the Ge atoms reaching the Si/SiO_2 interface reduce SiO_2 with deformation of volatile GeO units and thus the Si–Si bonds are generated. In fact, TEM images for the sample annealed at 1000°C show the shrinkage of the stacking dot structure due to reduction of SiO_2 in among dots (Figure 17.3). Also, the in-plane current–voltage characteristic (Figure 17.4) measured with co-planar electrodes implies an increase in the current density under 15 V/cm up to 18.5 nA/cm^2 for the sample annealed at 1000°C, while for the 800°C annealed sample the current density is 2.9 nA/cm^2. This *I–V* characteristic confirms the improvement of Si crystallinity through the formation of Si–Si bonds after high annealing temperature, and the contact properties of the metal electrodes is negligible due to the distance between electrode is far away compared with the dot size.

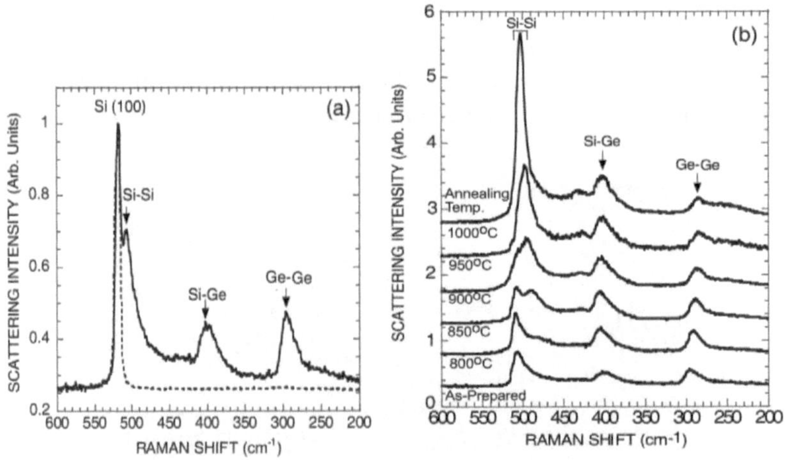

Figure 17.2 Raman spectra of a five-stacked structure of the Si dots with Ge core (a) as-prepared and (b) after annealing at various temperatures in N_2 ambient. In (a), the signals due to the TO phonon mode in the Si(100) substrate (dashed curve) overlap with those of the dots. For spectra in (b), the signals originating from Si(100) was subtracted from the as-measured spectrum.

The intermixing between the Si clad and the Ge core was also characterized using the XPS analysis of a single layer of dots formed on 10 nm-thick SiO_2. Figure 17.5 shows the XPS spectra of Si2p and Ge3d spectra for annealed samples as a function of annealing temperatures. With increasing annealing temperature, Si^{0+} signals

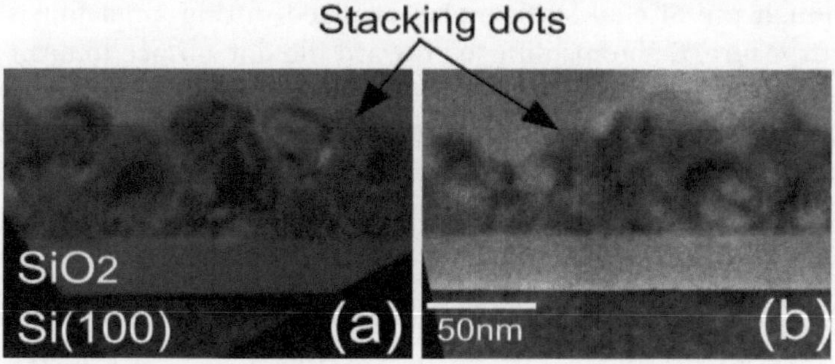

Figure 17.3 Cross-sectional HR-TEM images of samples with a dot-stacked structure annealed at (a) 900°C and (b) 1000°C.

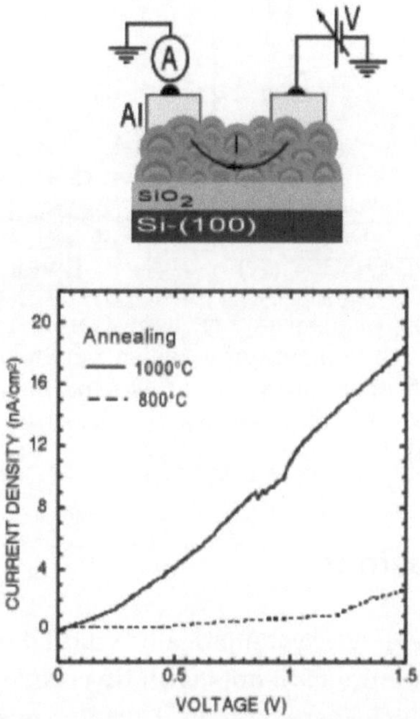

Figure 17.4 Schematic measurement set-up of in-plane *I–V* characteristics of stacked structure of Si dots with Ge cores after annealed at 800°C and 1000°C (a) and plot of in-plane *I–V* characteristics for each case.

originating from the Si clad decrease gradually, and the Si^{4+} signals from the underlying SiO_2 and surface oxide on the dot surface and Ge^{0+} signals from Ge core increase slightly. The result indicates that, although the Si clad surface was oxidized during annealing after air exposure, Ge atoms diffused toward the dot surface to form the compositional mixing at Si/Ge interface. Notice that chemically shifted Ge3d signals assigned to Ge^{4+} in oxide increase until it becomes its maximum at an annealing temperature around 650°C. This indicates that thermal desorption of GeO_x becomes significant at temperatures higher that 650°C.

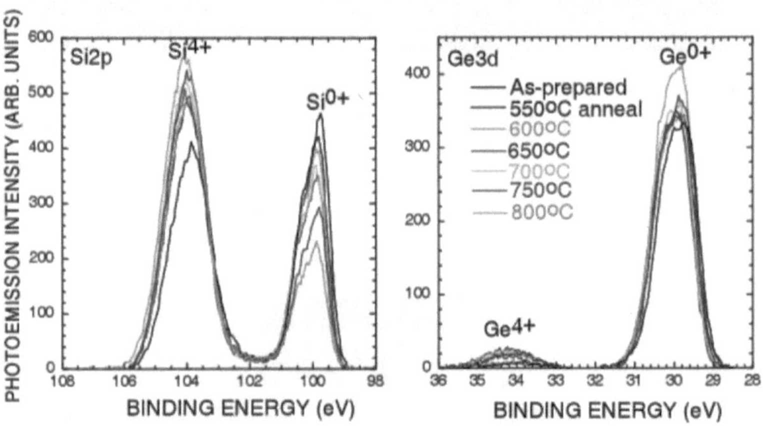

Figure 17.5 Si2p and Ge3d spectra for as-prepared and annealed samples at different temperatures, which were taken at photoemission take-off angle of 90°. For this experiment, a single layer of Si dots with Ge cores was prepared on 10 nm-thick SiO_2/Si(100).

17.4 Conclusions

In conclusion, we have systematically studied the compositional mixing and crystallinity of Si dots with Ge core (~20 nm in average dot diameter), which were prepared on thermally grown SiO_2 by highly selective LPCVD, as a function of annealing temperature in the range of 540–1000°C. Raman scattering spectra of multiply-stacked structures consisting of dots with Ge core and 2 nm-thick SiO_2 interlayers indicate the compositional mixing occurs partly at Si/Ge interface during the sample preparation, where the sample was heated up to 600°C. Also, 800°C annealing promotes Si–Ge alloying between Si clad and Ge core and degrades the crystallinity. By 950°C annealing, SiO_2 reduction with diffused Ge atoms to form volatile

GeO_x units becomes significant, resulting in the generation of Si–Si bonds. XPS analysis of annealed single layer structures confirms the incorporation of Ge atoms into Si clad from Ge core and the formation of GeO_x at the Si clad surface

Acknowledgments

The author would like to thank Prof. Miyazaki for inspiring discussion and Dr. H. Nakagawa, Dr. A. Ohta, and Dr. N. Kosku for their help in sample measurements.

References

[1] Vescan, L. (1998), Lateral ordering of Ge islands along facets, *J. Crystal Growth*, 194, 173–177.

[2] Schorer, R., Friess, E., Eberl, K., Abstreiter, G. (1991) Structural stability of short-period Si/Ge superlattices studies with Raman spectroscopy, *Phys. Rev. B*, 44(4), 1772–1781.

[3] Wan, J., Luo, Y. H., Jiang, Z. M., Jin, G., Liu, J. L., Kang, L., Wang, L., Liao, X. Z., Zou, J. (2001) Ge/Si interdiffusion in the GeSi dots and wetting layers, *J. Appl. Phys.*, 90(8), 4290–4292.

[4] Kamins, T.I., Medeiros-Ribeiro, G., Ohlberg, D.A.A., William, R.S. (1999) Evolution of Ge islands on Si(001) during annealing, *J. Appl. Phys.*, 85(2) 1159–1171.

[5] Miyazaki, S., Yamamoto, Y., Yoshida, E., Ikeda, M., Hirose, M. (2000) Control of self-assembling formation of nanometer silicon dots by low pressure chemical vapor deposition, *Thin Solid Films*, 369, 55–59.

[6] Darma, Y ., Takaoka, R., Murakami, H., Miyazaki, S. (2003) Self-assembling formation of SiQDs with Ge core by low-pressure chemical vapor deposition, *Nanotechnology*, 14(4), 413–415.

[7] Darma, Y., Takaoka, R., Murakami, H., Miyazaki, S. (2003) Formation of nanometer Si dots with Ge core by highly-selective low-pressure chemical vapor deposition, *Jpn. J. Appl. Phys.* 42(6B), 4129–4133.

Chapter 18

Application of TEM Nano-Beam Diffraction Strain Analysis in Advanced Semiconductor Devices

Zhu Jie,* Zhou Yong Kai, Du An Yan, and Hua You Nan
GLOBALFOUNDRIES, 60 Woodlands Industrial Park D, Street 2, 738406 Singapore
*zhujie@globalfoundries.com

18.1 Introduction

With the continuous narrowing of a transistor's gate length to deep sub-micron regime, strain engineering has become a predominant strategy to enhance charge carriers' mobility in MOSFET devices. Mobility enhancement of strained-Si is mainly due to decrease in the average effective mass of carriers resulting from stress-induced conduction sub-bands energy shift (n-type Si) and valence band wrapping (p-type Si) [1]. In addition, reduction in scattering rate due to conduction/valence sub-bands splitting further increases carriers' mobility. It has been demonstrated that the drive current of strained-Si pMOS and nMOS transistors can be potentially enhanced by ~4.5× and ~2×, respectively [2–3]. Various strain engineering

Advances in Nanodevices and Nanofabrication: Selected Publications from Symposium of Nanodevices and Nanofabrication in ICMAT2011
Edited by Qing Zhang, W. I. Milne, and Jianping Zou

ISBN 978-981-4364-54-6 (Hardcover), 978-981-4364-55-3 (eBook)
www.panstanford.com

techniques have been developed, such as nitride stress liner (NSL), stress proximity technique (SPT), stress memorization technique (SMT), diamond-like carbon (DLC) stressor, and embedded SiGe/SiC structures [4]. For example, drive current as high as 714 μA/μm (V_{dd} = 1 V, I_{off} = 100 nA/μm) was reported by Yasutake *et al.*, who applied a two-step recessed SiGe source/drain structure combined with 1.8 GPa compressive nitride stress liner [5]. However, it should be noted that strain may become harmful if the level of strain is too high. The formation of strain-induced Si defects was observed in the channel of a 45 nm pMOSFET with ~2.1 GPa compressive stress along the channel direction [6]. Strain-induced Si defects can also be generated after high temperature (~1000°C) IC fabrication process due to mismatch between thermal expansions coefficients (CTE) of different materials, such as after thermal oxidation and post-implantation spike annealing processes. Strain level must be carefully controlled to avoid defect formation and reduce the potential reliability issue. As the transistor size reduces, strain variation becomes much localized and depends strongly on the layout of integrated circuits. Therefore, accurate and precise strain measurement is important in monitoring process control and improving device performance.

Many techniques have been developed for strain measurement including Raman spectroscopy, X-Ray diffraction, convergent beam electron diffraction (CBED), high-resolution TEM imaging (HRTEM), electron holography and nano-beam electron diffraction (NBED). Raman spectroscopy and X-Ray diffraction have high accuracy; however the spatial resolution is limited to sub-micron range at best. Therefore, these techniques are not suitable for local strain measurement of patterned ICs. HRTEM can be used to measure strain variation of ~1×10^{-3} with high spatial resolution [7]. However, HRTEM imaging requires very thin sample thickness which causes strain relaxation and sample bending. As a result, the measured strain is deviated from its true value. CBED is the method that probably provides the most sensitive measurement (~1×10^{-4}) among all TEM techniques [8]. High spatial resolution of CBED is obtained using highly convergent electron beam. Nevertheless, CBED is limited by the fact that it requires sample tilt away off the low-index zone axis of Si substrate, i.e., <110> or <100> direction.

Such tilt may cause structures to overlap and therefore unreliable result. Convergent beam also damages the sample after prolonged exposure time. Splitting of HOLZ lines makes CBED analysis even more difficult. Recently, dark-filed off-axis electron holography has been shown to provide sensitive strain measurement ($\sim 1 \times 10^{-3}$ variation) with nanometer spatial resolution [9]. In addition, this technique can give strain mapping of micron-scale field of view which is not attainable by CBED and HRTEM techniques. But, relatively complex experimental set-up (both sample preparation and TEM recording) and analysis of holograms make it less practical in industrial routine work. Among all TEM techniques, nano-beam electron diffraction (NBED) is a favorable method for routine strain measurement because of its high spatial resolution, high sensitivity, and relative ease of data collection and analysis. NBED was firstly applied in the strained-Si on SiGe-on-insulator (SGOI) structure and shows ~10 nm spatial resolution and high sensitivity (1×10^{-3}) [10]. Strain is measured by recording diffraction patterns with nearly parallel electron probe under scanning-TEM mode. From diffraction patterns, lattice spacing of different crystallography orientations can be obtained. Strain is then calculated by taking lattice spacing from unstrained Si substrate as reference. Unlike CBED, NBED prefers small convergence angle of electron probe in order to obtain small and sharp diffraction spots which can improve accuracy and precision of measurement. Later development using chromatic aberration corrector and energy filter further improves the measurement sensitivity (6×10^{-4}) and spatial resolution (2.7 nm) [11].

In this work, NBED technique was applied to investigate process-induced strain variation in Si substrate of advanced semiconductor devices. In the first study, NBED was used to identify the thermal process inducing large strain and defects at source/drain locations. Stress accumulation was suspected after several high-temperature manufacturing processes. It is necessary to quantitatively show the strain change after each suspected process. With strain measurement after different processes, it confirms that laser spike annealing after source/drain ion implantation is the major stress source. In the second case, we used NBED to show that there is ~100% strain enhancement in the channel area of embedded SiGe structure upon removal of poly gate. This result reveals that gate-last process is more advantageous as compared to the gate-first process from the strain engineering perspective.

18.2 Experiment

18.2.1 Sample Preparation

TEM samples used for NBED strain measurement were prepared using dual-beam FIB milling (FEI Helios 400S) at 30 kV ion beam voltage. *In situ* lift-out FIB technique was used to reduce sample surface contamination, which raises background noise of diffraction pattern. Sample thicknesses were controlled at ~400 nm as to reduce strain relaxation and bending effect. Before FIB milling, sample surface was coated with ~100–200 nm SiO_2 using precision etching and coating system (Gatan PECS) as protective layer. Plasma cleaning was performed before sample was loaded into TEM in order to reduce carbon contamination accumulated on sample surface.

18.2.2 NBED Analysis

NBED strain measurement was performed under scanning-TEM mode using FEI Titan 80-300 S-Twin TEM equipped with Schottky FEG source and imaging spherical aberration corrector. Nearly-parallel electron probe of small semi-convergence angle (~0.3 mrad) was used in order to obtain sharp and small-diameter diffraction spots. Z-contrast STEM image was taken with HAADF detector and series of diffraction patterns from a line scan were recorded on a CCD camera. Each electron diffraction pattern can be used to calculate strain at that location. To enhance the signal-to-noise ratio in diffraction pattern, Gatan Image Filter (GIF) was used to screen out partial inelastic scattered electrons at slit width of 10 eV. Electron probe size used in our experiment is estimated to be less than ~5 nm [12]. Figure 18.1 gives simplified electron diffraction geometry from which we can find approximate value of lattice parameter using the following equation:

$$R \cdot d = L \cdot \lambda \tag{18.1}$$

where, R is the distance measured from the center maxima of direct beam (O′) to the diffracted beam (G′) on the recorded diffraction pattern, d is the distance between parallel planes which give rise to the diffracted beam G′, and L and λ are the cameral length and wavelength of electron beam, respectively. Based on equation 18.1, it is easy to find out the strain at a strained location:

$$\varepsilon_i = (d_i - d_o)/d_o = (R_o - R_i)/R_i \qquad (18.2)$$

where, ε_i is the strain, d_0, R_0 and d_i, R_i are the values obtained from diffraction patterns corresponding to the reference and strained locations, respectively. The reference location used in this work was chosen on un-strained Si substrate ~2 μm below the Si surface. The direction of measured strain is determined by the diffraction spot vector selected for calculation. Clearly from equation 18.2, we only need to measure the distance R in order to know the strain value, since both L and λ are constants for a fixed experiment condition. The digital diffraction pattern allows us to obtain distance R using image processing software. Here, we applied a public domain software (Lispix) written by David at National Institute of Standards and Technology (NIST) which enables us to identify the center maxima of the diffraction spots [13]. Maximum image pixel size (2048 × 2048) is used for each digital diffraction pattern in order to reduce measurement error of R. The experimental error can be further reduced by using distance R between G′ and –G′ instead of between O′ and G′. To be confident with analysis result, we investigated strain measurements (20 points taken for each sample) on un-strained Si substrate and obtained measurement sensitivity of 1×10^{-3} (standard deviation).

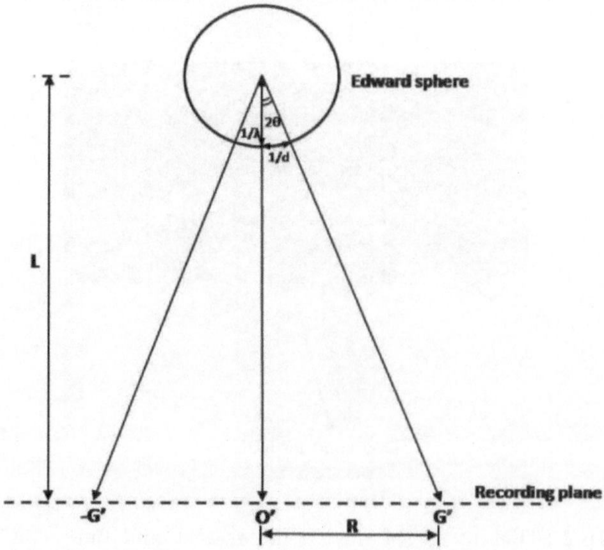

Figure 18.1 Schematic drawing of electron diffraction geometry.

18.3 Results and Discussion

18.3.1 Thermal Stress-Induced Si Defects

The device under evaluation suffered from high current leakage due to formation of Si defects at MOSFETs source/drain area. Defects were observed after high-temperature laser spike annealing (LSA) for activating source/drain dopants (Figure 18.2). These defects are suspected to be caused by the strain that could be introduced either during LSA or accumulated through previous thermal processes. Based on the design of experiment (DOE) results from manufacturing line, strain source could be from two different thermal processes: gate oxide growth and/or laser spike annealing after source/drain implantation. LSA is a common technique applied in advanced devices to drive dopant diffusion and anneal Si damage. Careful control of LSA parameters is critical in minimizing harmful strain such as LSA cycles and laser probe scanning direction. Gate oxidation is processed before source/drain LSA and is also a process performed under high temperature environment. Large strain can be introduced to the isolated Si substrate from surrounding shallow trench isolation (STI) oxide due to the difference in their thermal

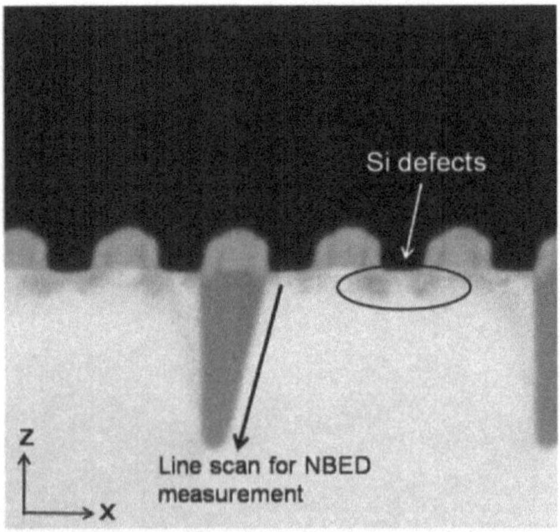

Figure 18.2 STEM image of device under test and line scan position. Distorted Si/STI structure is believed due to relatively large stress during thermal processes.

expansion coefficient. Either compressive or tensile strain can be present in Si depending on the mechanisms of strain formation, such as thermal mismatch, volume change during oxidation and density change in oxide densification. For example, compressive strain is generated inside Si substrate during thermal oxidation due to conversation of Si to SiO_2 with volume increase. To verify the strain source, NBED was used to measure and compare local strain variation after gate oxidation and LSA separately. The strain measurement location was chosen to be near to Si/STI oxide interface (as indicated by arrowed line in Figure 18.2) since the strain level is usually the maximum at interface and decreases as moving into Si horizontally.

The measurement tried to avoid Si defect region because it can lead to uncertainty in the measurement consistency. Diffraction patterns of 10 points were taken on each line scanned from top to bottom and strain was calculated along transistor channel direction (x-axis) using (022) and (−022) diffraction spots. Figures 18.3 and 18.4 show the STEM images, selected diffraction patterns and strain plots measured on two sides of Si active from sample processed after gate oxidation. From STEM images, Si defects were not formed after gate oxidation. Strain plots also indicate that there is no obvious strain increase after gate oxidation since all the values fall into standard deviation of measurement sensitivity.

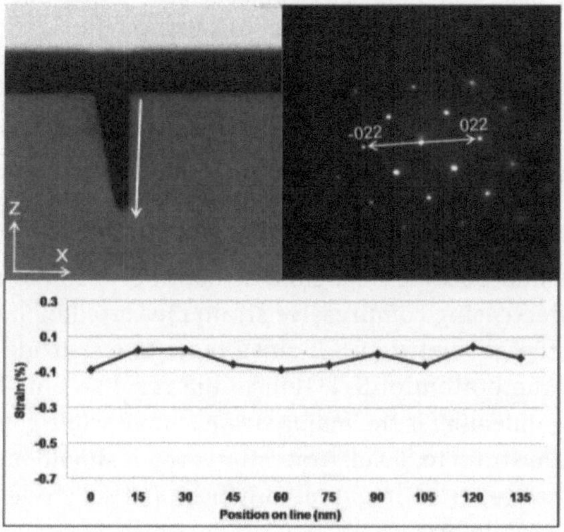

Figure 18.3 Line-scan strain measurement near Si/STI interface on the sample processed after gate oxidation.

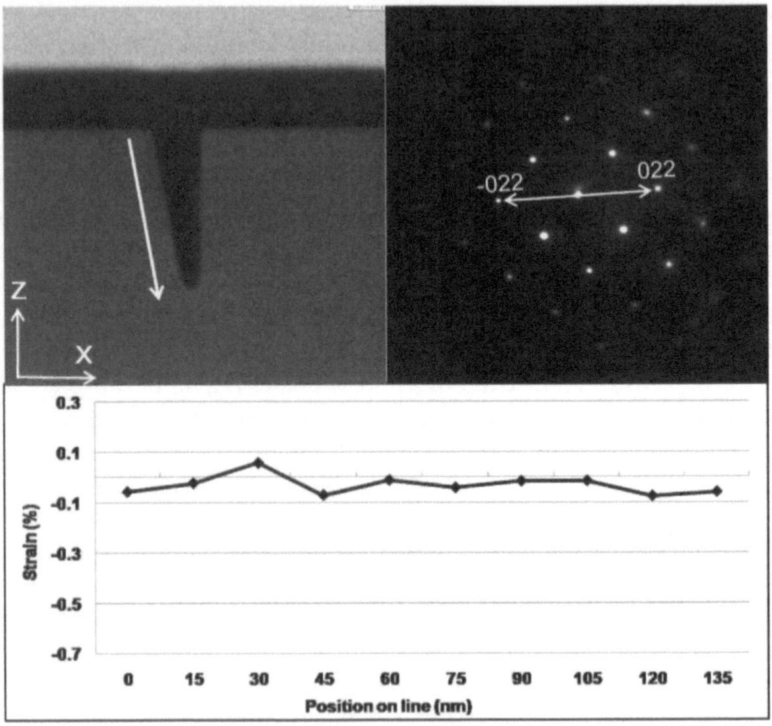

Figure 18.4 Line-scan strain measurement near Si/STI interface on the sample processed after gate oxidation.

For comparison, Figures 18.5 and 18.6 give the NBED strain results after laser spike annealing. Si defects can be observed at transistor source/drain area. From strain plots, relatively large compressive strain was detected at the beginning of the lines in both sides of Si active. Compressive strain decreases as measurement moves downward and become negligible at depth of ~45–50 nm on line scan. The decreasing compressive strain measured in Si is probably related to the geometry of STI since there is less oxide volume as moving to the bottom of STI. Hence, the results demonstrate that laser spike annealing is the major strain source while gate oxidation induces little strain to Si substrate. However, it should be noted that the compressive strain might already been partially released due to formation of Si defects.

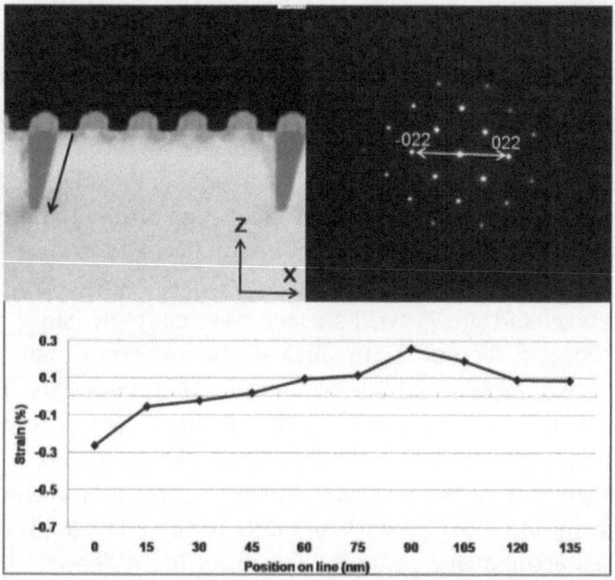

Figure 18.5 Line-scan strain measurement near Si/STI interface on the sample processed after LSA.

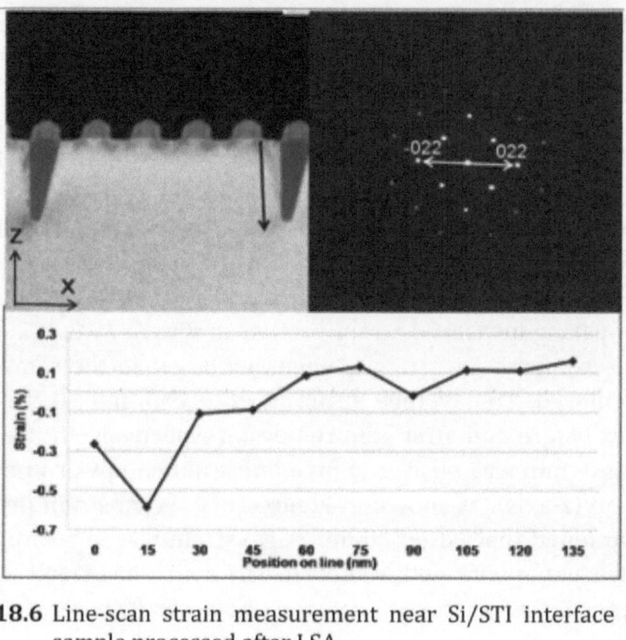

Figure 18.6 Line-scan strain measurement near Si/STI interface on the sample processed after LSA.

18.3.2 Thermal Stress-Induced Si Defects

In this case, NBED was applied to investigate strain profile in pMOSFETs with embedded-SiGe (eSiGe) grown at source/drain junction. Epitaxial eSiGe can induce large compressive strain in channel direction due to lattice mismatch between SiGe and Si. Optimization of device performance through strain engineering is controlled by many such processing parameters as Ge concentration, eSiGe thickness, shape and structure of eSiGe, growth/annealing temperature, and pre-growth surface treatment, etc. Since channel is very close to the substrate surface, channel strain may become sensitive to surface structure modification/materials removal during manufacturing process. In the advanced devices employing high-k metal gate structures, poly-Si gate electrodes are still used in the beginning of the process. But poly gates will eventually be etched away and replaced with metals in the gate-late approach. The step of gate removal is one major processing difference between the gate-late and gate-first processes. The removal of poly gate may induce channel strain change due to less restriction on top of channel surface. Therefore, it is important to study channel strain variation upon poly gates removal. Figure 18.7a shows TEM image of sample used in this study which was processed after eSiGe growth and patterned with poly gate electrode. Figure 18.7b shows the sample profile after poly gate removal. Gate electrodes were removed by scratching sample surface with cotton stick and the region showing complete SiGe structure is selected for analysis. Firstly, strain was measured on a line scanning in channel direction (x-axis) and located at ~10 nm below Si substrate surface. Strain was calculated along channel direction using (022) and (−022) diffraction spots. By comparing measured strain profiles in Figure 18.8, it is evident that the compressive strain is generally larger in the sample with gate removed. The average strain is ~(−0.4%) and −0.7% for the samples before and after gate removal, respectively. In the second analysis, strain was measured on a line scanned upward in vertical direction (z-axis), as shown in Figure 18.9. From strain profiles, it is again noted that larger compressive strain was measured in the line for sample with gate removed. The strain has small variation at deeper Si region in both samples; however it increases sharply as the measurement moves upward to a depth of ~30 nm for the sample without poly gate. The strain is relatively uniform along line scanning for sample with poly gate. The above results indicate that there is ~100% strain enhancement (as measured at depth of 10 nm in Figure 18.9) in the channel area after poly gate removal. Strain increase is also observed at deeper Si area but larger increment is near Si surface. These results reveal that eSiGe strain engineering maybe more effective in the gate-late process where poly-Si removal is necessary.

Figure 18.7 STEM images of (a) eSiGe structure with poly gate and (b) after removal of poly gate.

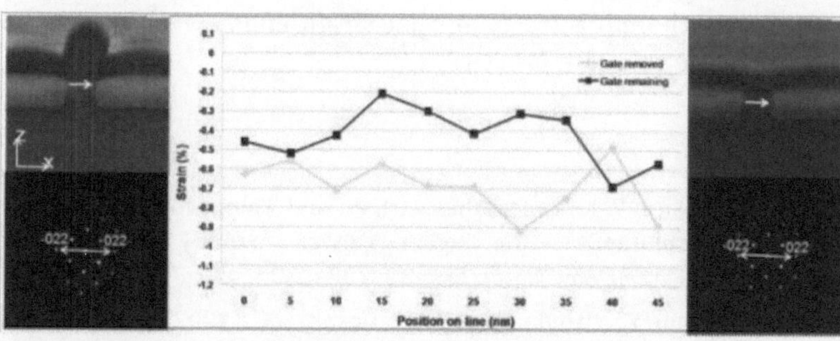

Figure 18.8 Shows strain profiles measured on horizontal line-scans in channel area of eSiGe structures before/after poly gate removal. The strain direction corresponds to channel direction.

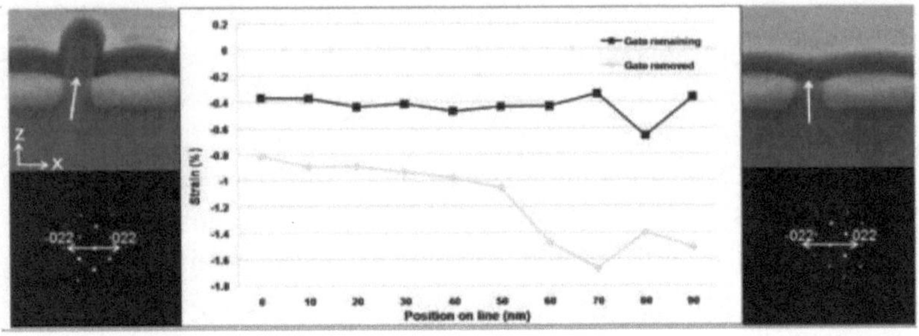

Figure 18.9 Shows strain profiles measured on vertical line-scans in channel area of eSiGe structures before/after poly gate removal. The strain direction corresponds to channel direction.

18.4 Conclusion

In summary, we have applied NBED technique to measure local strain variation resulted from different manufacturing processes and structure change. Using NBED, it is able to monitor strain increase after each processing step and help verify strain source. In the first case, we measured strain after thermal oxidation and laser spike annealing which are suspected to cause Si defects in substrate. Through analysis comparison, it confirms that spike annealing after source/drain implantation is the major strain source. In the second case, we demonstrated that channel strain in eSiGe structure is sensitive to surface structure change. After removal of poly gate, there is ~100% increment of strain due to less restriction on channel surface and therefore implying that strain engineering maybe for favorite in the gate-late process.

References

1. Chu, M., Sun, Y. K., Umamaheswari, A., and Thompson, S. E. (2009), Strain: a solution for higher carrier mobility in nanoscale MOSFETs, *Annu. Rev. Mater. Res.*, **39**, 203–239.
2. Sun, G., Sun, Y., Nishida, T., and Thompson, S. E. (2007), Hole mobility in silicon inversion layers: stress and surface orientation, *J. Appl. Phys.*, **102**, 084501.
3. Suthram, S., Ziegert, J. C., Nishida, T., and Thompson, S. E. (2007), Piezoresistance coefficients of (100) silicon nMOSFETs measured at

low and high (~1.5GPa) channel stress, *IEEE Electron. Dev. Lett.*, **28**, 58–61.

4. Sun, Y. K., Thompson, S. E., and Nishida, T. (2010), *Strain Effect in Semiconductors: Theory and Device Applications* (Springer, USA), Chapter 7 "Strain in electron devices," pp. 235–266.
5. Yasutake, N., Zuma, A., Ishida, T., Kusunoki, N., Mori, S., Itokawa, H., Mizushima, I., Okamoto, S., Morooka, T., Aoki, N., Kawanaka, S., Inaba, S., and Toyoshima, Y. (2007), Record-high performance 32nm node pMOSFET with advanced two-step recessed SiGe-S/D and stress liner technology, *Symposium on VLSI Technology*, pp. 48–49.
6. Wong, B. P., Zach, F., Moroz, V., Mittal, A., Starr, G. W., and Kahng, A. (2009), *Nano-CMOS Design for Manufacturability: Robust Circuit and Physical Design for Sub-65 nm Technology Nodes* (John Wiley & Sons, New Jersey), Chapter 3 "Interaction of layout with transistor performance and stress engineering techniques," pp. 146–147.
7. Hüe, F., Hÿtch, M., Bender, H., Houdellier, F., and Claverie, A. (2008), *Phys. Rev. Lett.*, **100**, 156602.
8. Zhang, P., Istratov, A. A., Weber, E. R., Kisielowski, C., He, H. F., Nelson, C., and Spence, J. C. H. (2006), *Appl. Phys. Lett.*, **89**, 161907.
9. Hÿtch, M., Houdellier, F., Hüe, F., and Snoeck, E. (2008), Nanoscale holographic interferometry for strain measurements in electronic devices, *Nature*, **453**, 1086–1089.
10. Usuda, K., Numata, T., and Takagi, S. (2005), Strain evaluation of strained-Si layers on SiGe by the nano-beam electron diffraction (NBD) method, *Mater. Sci. Semicond. Proc.*, **8**, 155–159.
11. Béché, A., Clement, L., and Rouvière, J. L. (2010), Improved accuracy in nano beam electron diffraction, *JPCS*, **209**, 012063.
12. Sourty, E., Stanley, J., and Freitag, B. (2009), *IEEE Proceedings of 16th IPFA*, pp. 479–484.
13. http://www.nist.gov/lispix/LxDoc/home.html.

Chapter 19

Electron Mobility and Persistent Photoconductivity in Quantum Wells $In_{0.52}Al_{0.48}As/In_{0.53}Ga_{0.47}As/In_{0.52}Al_{0.48}As$ on InP Substrate

Vladimir A. Kulbachinskii,* Roman A. Lunin, Natalia A. Yuzeeva, Galib B. Galiev, Ivan S. Vasilievskii, and Eugene A. Klimov
Moscow State University (M.V. Lomonosov), Physics Faculty,
Low Temperature Physics and Superconductivity Department,
Moscow, 119991 GSP-1, Russia
*kulb@mig.phys.msu.ru

Isomorphic $In_{0.52}Al_{0.48}As/In_{0.53}Ga_{0.47}As/In_{0.52}Al_{0.48}As$ quantum well structures on InP substrate were grown by molecular beam epitaxy (MBE). We investigated the electron transport properties and mobility enhancement in the structures by changing of doping level, the width d of quantum well $In_{0.53}Ga_{0.47}As$ or by illumination using light with $\lambda = 668$ nm. Persistent photoconductivity was observed in all samples due to spatial separation of carriers. We used Shubnikov-de Haas (SdH) effect to analyze subband electron concentration and mobility.

Advances in Nanodevices and Nanofabrication: Selected Publications from Symposium of Nanodevices and Nanofabrication in ICMAT2011
Edited by Qing Zhang, W. I. Milne, and Jianping Zou

ISBN 978-981-4364-54-6 (Hardcover), 978-981-4364-55-3 (eBook)
www.panstanford.com

19.1 Introduction

There is a significant interest to InAlAs/InGaAs heterostructures on InP substrate. One of the reasons is the advantage of such structures. They allow using the wider frequency range and having lower electronic noise. The possibility to increase the InAs content in the InGaAs layer up to 70% and higher gives an opportunity to increase mobility, electron concentration in the channel and the drift velocity of electrons. Isomorphic or pseudomorphic InGaAs structures on InP substrate may be used for different applications. In one of the first [1], matched heterostructures InAlAs/InGaAs/InAlAs on InP substrate were grown. The electron mobility and concentration were investigated as a function of the width of spacer and quantum well InGaAs width *d*. Later for HEMT isomorphic heterostructures on InP substrate with channel d = 20 nm and spacer 2 nm were used [2]. It is known also that inserted InAs layers and composite conducting channel with shortness of the gate improved the UHF characteristics of transistors [3]. In this research, we investigated the electron transport properties and mobility enhancement in the ternary structures $In_{0.52}Al_{0.48}As/In_{0.53}Ga_{0.47}As/In_{0.52}Al_{0.48}As$ on InP substrate by changing of doping level, the width *d* of isomorphic quantum well $In_{0.53}Ga_{0.47}As$ or by illumination using light with λ = 668 nm. We used SdH effect to analyze subband electron concentration and mobility and their dependence on doping level, quantum well width and illumination.

19.2 Experimental

The samples were grown by the MBE on InP (100) substrates. All samples have $In_{0.52}Al_{0.48}As$ buffer matched (isomorphic) to InP. Samples have one-side delta-doping by Si. The spacer thickness was 4.3–6.0 nm. Cap layer was undoped $In_{0.53}Ga_{0.47}As$. Schematic diagram of the $In_{0.52}Al_{0.48}As/In_xGa_{1-x}As/In_{0.52}Al_{0.48}As$ structure on InP substrate is shown in Figure 19.1. Temperature dependence of the resistance and Hall effect were measured in the temperature interval 4.2 K < T < 300 K. At T = 4.2, K SdH effect was investigated in magnetic field provided by a superconducting solenoid up to 6 T. We found that mobility depends on doping and width of quantum well. Some experimental parameters of samples at T = 4.2 K are listed in Table 19.1.

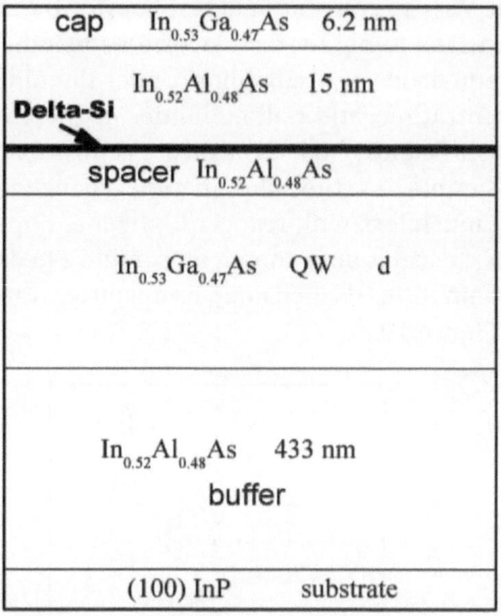

Figure 19.1 Schematic diagram of the $In_{0.52}Al_{0.48}As/In_{0.53}Ga_{0.47}As/In_{0.52}Al_{0.48}As$ structure on InP substrate.

Table 19.1 Some parameters of samples: quantum well width *d*, Hall concentration n_H and mobility μ_H at 4.2 K, electron concentration determined from SdH effect in first n_1 and second n_2 subbands

Sample	*d*, nm	n_H, 10^{12} cm^{-2} (dark)	μ_H cm^2/(Vs) (dark)	μ_H cm^2/(Vs) (illuminated)	$n_1(n_2)$ 10^{12} cm^{-2} (dark)	$n_1(n_2)$ 10^{12} cm^{-2} (illuminated)
1	26	3.2	40000	41000	2.49(0.70)	2.50(0.74)
2	18.5	2.6	45800	46900	2.00(0.59)	2.20 (0.7)
3	16	1.95	53500	60000	1.67(0.26)	1.87 (0.53)
4	14.5	1.6	45000	52400	1.55 (—)	2.08 (—)

19.3 Results and Discussion

We observed two frequencies corresponding to two filled subbands in SdH effect in all samples except sample 4 due to much less doping of this sample. We use conventional fast Fourier transform (FFT) of the SdH oscillations to obtain the electron subband concentrations n_i from the frequencies of oscillations. Under illumination subband electron concentrations and Hall mobilities μ_H increased (see Table 19.1). When a sufficiently high electron concentration was produced by doping (sample 1) the illumination changed the electron concentration much less with respect to the less doped samples. In Figure 19.2, SdH oscillation is shown for sample 2 in dark (solid line) and under illumination (dashed line). Fast Fourier transform spectra are shown in Figure 19.3.

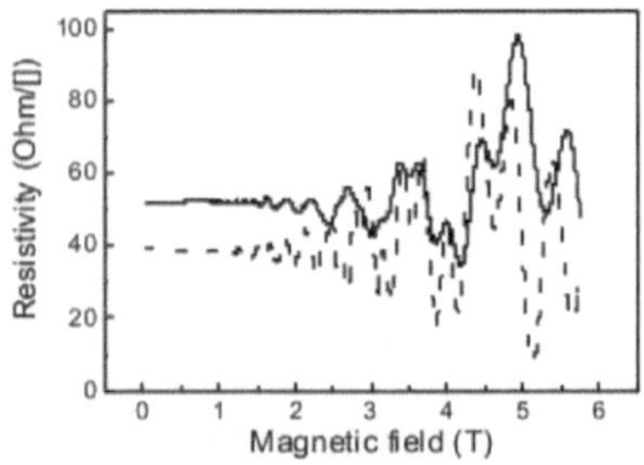

Figure 19.2 Magnetoresistance oscillation in sample 2 at $T = 4.2$ K in dark (solid) and under illumination (dashed).

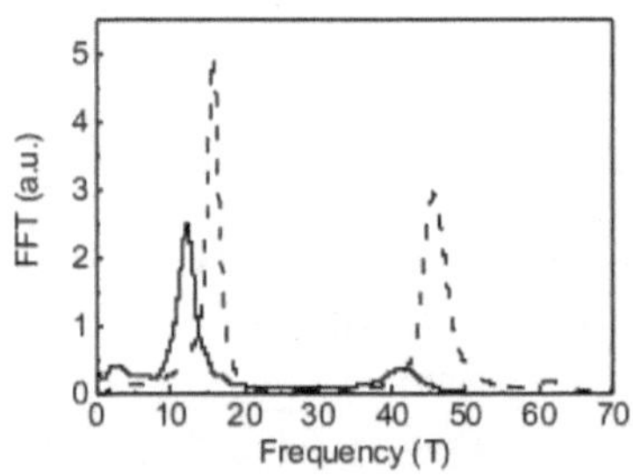

Figure 19.3 Fourier spectra of SdH oscillation for sample 2 in dark (solid) under illumination (dashed).

Persistent photoconductivity was observed in all samples. The typical dependences of the resistance on temperature and illumination are shown in Figure 19.4 for two different samples. The samples were illuminated at $T = 4.2$ K to achieve the saturation of its resistance, then the illumination was switched off and the temperature dependence of the resistance was measured with the rate 5 K/min. At $T > 160$–170 K, the difference between resistances measured in the dark and after illumination became less than the error of measurements.

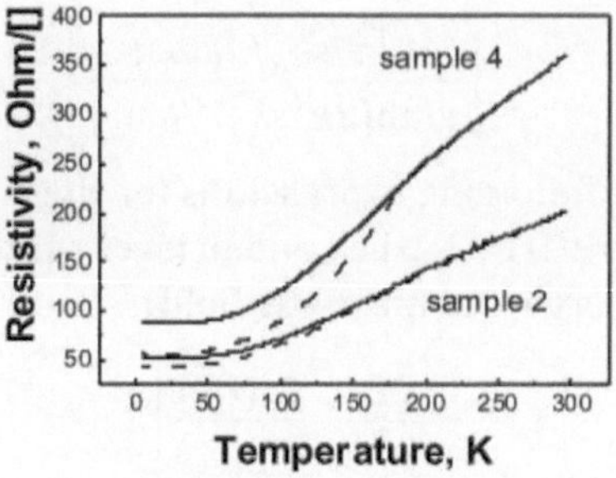

Figure 19.4 Temperature dependence of the sheet resistance for sample 2 and 4 in dark (solid line) and after illumination at $T = 4.2\ K$ (dashed line).

The effect of persistent photoconductivity is related to the spatial separation of carriers. This fact is confirmed by the logarithmic dependence of the photoconductivity relaxation on time in the initial time interval $\Delta\sigma = \sigma\ (0) - \sigma(t) = A\ \ln(1 + t/\tau)$ [4,5] with the value of τ in the range 20–30 s. Relaxation and fitting of conductivity for sample 3 is shown in Figure 19.5.

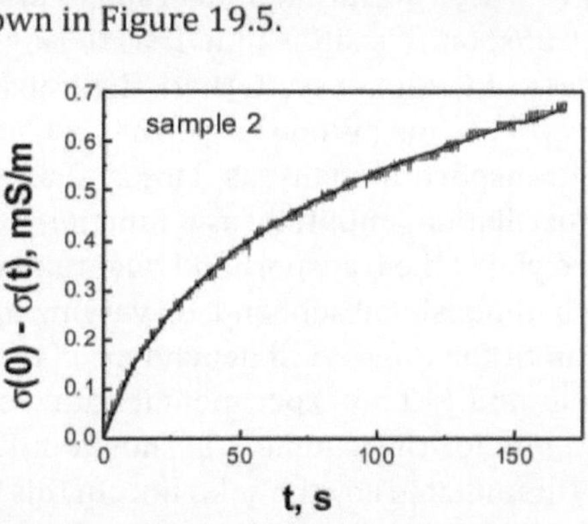

Figure 19.5 Relaxation of resistivity of sample 2 at T = 4.2 K in dark after illumination. Points are experimental data; solid line is fitting.

Reliable electron densities and mobilities can be derived from SdH oscillations, for which each subband has an oscillation with its own period. The part of the density of states oscillating in magnetic field, Δg, can be expressed as [4]

$$\frac{\Delta g(\varepsilon_f)}{g_0} = \sum_{s=1}^{\infty} \exp\left(\frac{\pi s}{\pi_q B}\right) \cos\left[\frac{2\pi s(E_F - E_i)}{\hbar\omega_c} - s\pi\right] \left[\frac{(2\pi^2 s k_B T/\hbar\omega_c)}{\sinh(2\pi^2 s k_B T/\hbar\omega_c)}\right], \tag{19.1}$$

which leads to the following expressions for the conductivity tensor components in the 2-D case (the Landau level width is assumed to be independent of energy and magnetic field):

$$\sigma_{xx} = \frac{en_s\mu_t}{1+\mu_t^2B^2}\left[1+\frac{2\mu_t^2B^2}{1+\mu_t^2B^2}\frac{\Delta g(\varepsilon_F)}{g_0}\right], \tag{19.2}$$

$$\sigma_{xy} = -\frac{en_s\mu_t^2B}{1+\mu_t^2B^2}\left[1-\frac{3\mu_t^2B^2+1}{\mu_t^2B^2(1+\mu_t^2B^2)}\frac{\Delta g(\varepsilon_F)}{g_0}\right], \tag{19.3}$$

where μ_t is the transport mobility at $B = 0$, g_0 is the density of states at zero magnetic field B, μ_q is the quantum mobility, N_s is the electron density, and e is the absolute value of the electron charge. The oscillation frequency f in the reciprocal magnetic field determines the two-dimensional electron density: $N_s = ef/\pi\hbar$, and Fermi energy is $E_F = (\pi\hbar^2/m^*)N_s$. Scattering through small angles makes a negligible contribution to transport relaxation time. If scattering is large at small angles (in the case of Coulomb scattering), the transport relaxation time may be an order of magnitude larger than quantum one.

Hence the transport mobility is larger than quantum one. Analysis of the oscillation amplitude as a function of magnetic field and temperature yields the transport and quantum mobilities of 2D electrons in each dimensional subband. By varying μ_t and μ_q in each subband, one can fit the theoretical dependence (1–3) of oscillation on magnetic field and FFT to experimental data. For example, the fit to measurements for the sample 3 is shown in Figures 19.6 (in dark) and 19.7 (illuminated) by the solid line. In this fit, we used the optimization technique [4]. The resulting transport and quantum mobilities are listed in Table 19.2. In illuminated samples, there is a tendency to higher mobilities (both transport and quantum) in both filled subbands. The transport mobility is larger than the quantum mobility. The transport mobility can be derived using the kinetic equation and describing the impurity scattering in the Born approximation. The scattering theory was generalized to the case of several filled subbands in Ref. [6] (see also Refs. [4,7]). When several dimensional subbands are filled, the transport relaxation times $\tau_{t,i}$ are derived from the linear equation system

$$P_i(E)\tau_{t,i}(E) - \sum_{i \neq i'} P_{ii'}(E)\tau_{t,i}(E) = 1, \tag{19.4}$$

where the coefficients $P_{ii'}(E)$ are the probabilities for the respective intersubband transitions. The polarization component of the dielectric function we took from the Ref. [8]. The scheme for calculating transport relaxation times in the subbands described in [4]. The quantum lifetime $\tau_{q,i}$ at the Fermi level is obtained by adding all the scattering probabilities with equal weights [9]

$$\frac{1}{\tau_{q,i}} = \frac{m^*}{\pi\hbar^3}\sum_{i'}\int_0^{\pi} d\varphi \left|\tilde{V}_{ii'}(q')\right|^2, \mu_{q,i} = \frac{e}{m^*}\tau_{q,i} \tag{19.5}$$

The quantum and transport mobilities in sample 3 calculated by this method are listed in Table 19.3. We also show in Table 19.3 calculated and experimental subband electron concentrations N_s. The dependence of the mobility on the subband number is determined by several competing factors. On one hand, the Fermi momentum is lower in the higher subbands, which leads, according to the well-known properties of Coulomb scattering, to a lower mobility.

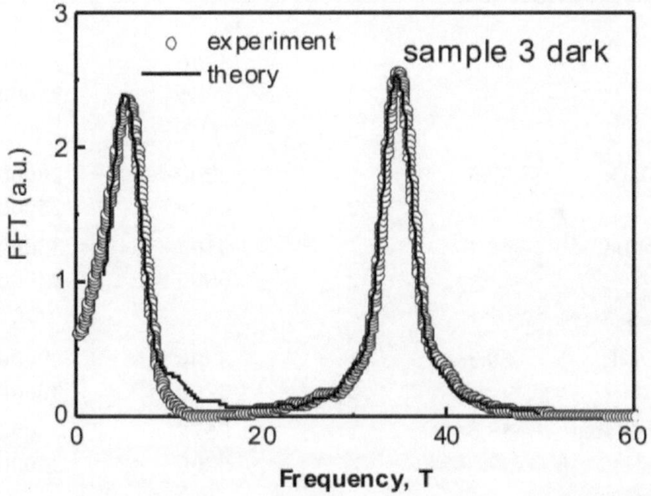

Figure 19.6 FFT of SdH oscillation for sample 3 in dark. Point are experimental data, solid line is fitting according to equations 19.1–19.3.

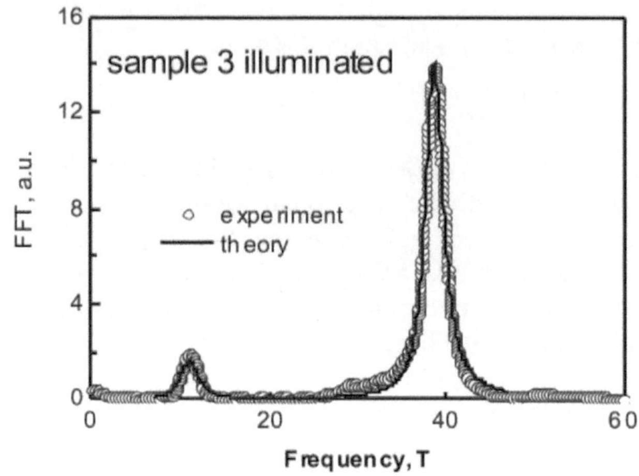

Figure 19.7 FFT of SdH oscillation for illuminated sample 3. Point are experimental data, solid line is fitting according to equations 19.1–19.3.

Table 19.2 Sheet resistivity ρ, quantum μ_q, and transport μ_t mobilities at T = 4.2 K

Sample	ρ, Ω/ρ	Subband #	μ_q, cm^2/Vs	μ_t, cm^2/Vs
1				
	47.36	1	4300	23000
		2	8300	21000
2				
Dark	52.5	1	2200	28000
		2	4400	25000
Illuminated	44.4	1	3500	35000
		2	5000	32000
3				
Dark	59.9	1	3200	35000
		2	3200	30000
Illuminated	43.6	1	5070	44000
		2	5200	36000
4				
Dark	88.69	1	2100	22000
Illuminated	57.3	1	2700	40000

Table 19.3 Quantum μ_q and transport μ_t mobilities, subband electron concentrations at $T = 4.2$ K

Sample 3	Subband	N_{SdH}, $10^{12}\cdot cm^{-2}$ experiment	N_S, $10^{12}\cdot cm^{-2}$ theory	μ_q, cm^2/Vs	μ_t,cm^2/Vs
Dark	2	0.26	0.33	12300	36700
	1	1.67	1.60	20000	43100
Illuminated	3	—	0.45	900	2300
	2	0.53	0.71	1200	3400
	1	1.87	1.69	31500	48600

On the other hand, the width of the electron localization region is larger in the higher subbands, i.e., the average separation between impurities and electrons is larger, which should result in a higher mobility. Our numerical calculations (see Table 19.3) indicate that the electron mobility decreases with the subband number. The mobility in the second subband is less because the wave function spread to the delta-layer where charged impurity scattering is high. The calculated transport mobility (see Table 19.2) is much higher than quantum mobility. The calculations and measurements of mobilities are in a good agreement, and the small discrepancies between the calculations and SdH measurements are, most probably, due to errors in the impurity distribution (which is important for the lower subbands) and in the calculated energy levels and wave functions (which affects mostly the mobility in higher subbands).

19.4 Conclusion

We investigated the electron transport properties and mobility enhancement in the isomorphic QW structures on InP substrate by changing of doping level and the width d of quantum well $In_{0.53}Ga_{0.47}As$. The maximal mobility was observed for QW width d = 16 nm. Persistent photoconductivity was observed in all samples due to spatial separation of carriers. Our numerical calculations indicate that the electron mobility decreases with the subband number. The mobility in the second subband is less because the wave function spread to the delta-layer where charged impurity scattering is high.

The calculated transport mobility is much higher than quantum mobility. The calculations and measurements of mobilities are in a good agreement.

References

1. Brown, A.S., Mishra, U.K., Henige, J.A., Delaney, M.J. (1988). The impact of epitaxial layer design and quality on GaInAs/AlInAs high-electron-mobility transistor performance, *J. Vac. Sci. Technol.* **B6**(2), 678–680.
2. K. Higuchi, H. Uchiyama, T. Shiota, M. Kudo, T. Mishima (1997). Selective wet-etching of InGaAs on InAlAs using adipic acid and its application to InAlAs/InGaAs HEMTs, *Semicond. Sci. Technol.* **12**, 1997, 475–480.
3. Malmkvist, Mikael, Wang, S, Grahn, Jan V. (2008). Epitaxial optimization of 130-nm gate-length InGaAs/InAlAs/InP HEMTs for high-frequency Applications, *IEEE Trans. Electron Devices* **55**(1), 268–275.
4. Kulbachinskii, V.A., Lunin, R.A., Kytin, V.A. (1996). Electron mobilities in dimensional subbands of combinatively doped GaAs/GaAlAs heterojunctions with high density of 2D electrons, *JETP* **93**, 1517–1532.
5. Queisser, H.J., Theodorou, D.E. (1986). Decay of persistent photoconductivity in semiconductors, *Phys. Rev. B* **33**, 4027–4033.
6. Siggia, E.D., Kwok, P.C. (1070). Properties of electrons in semiconductor inversion layers with many occupied electric subbands. I. Screening and impurity scattering, *Phys. Rev. B* **2**, 1024–1036.
7. Coleridge, P.T., (1991). Small-angle scattering in two-dimensional electron gases, *Phys. Rev. B* **44**, 3793–3801.
8. Inoue, K., Matsuno, T., (1993). Electron mobilities in modulation-doped $Al_xGa_{1-x}As$/GaAs and pseudomorphic $Al_xGa_{1-x}As$/$InyGa_{1-y}As$ quantum-well structures, *Phys. Rev. B.* **47**, 3771–3778.
9. Fletcher R., Zaremba, E., D'Iorio, M., Foxon C.T., Harris J.J., (1990). Persistent photoconductivity and two-band effects in GaAs/$Al_xGa_{1-x}As$ heterojunctions, *Phys. Rev. B* **41**, 10649–10666.

Index

advanced semiconductor devices 259, 261
AFM (atomic force microscope) 19–20, 39, 77, 109, 111, 164, 169, 194
Ag hole arrays 223–24, 227, 229
Ag holes 221, 224–28
Ag nanohole arrays 223, 228–29
amorphous carbon 68, 73, 88, 138
anisotropic etching 23–24, 31
annealing 64, 106–7, 109, 112–15, 251, 253–54, 256–57
annealing temperatures 51, 106–7, 114, 256
armchair segments 120–22, 124–26
atomic force microscope *see* AFM
axicon 196–97

ballast resistances 59, 61
bandgap 154–55, 157, 159, 182–83
beamlets 190–92, 194, 196–97
biasing electrodes 59
binding energy 239, 241–42
bird's beak structures 18–19
building units *see* BUs
BUs (building units) 62, 81–82, 87

carbides 79, 171
carbon 50–51, 71–72, 77, 79–80, 88, 90, 94–97, 120, 127–28, 136, 149–50, 155, 169, 187
carbon atoms, defected 131
carbon nanostructures 86, 88, 96
carbon nanotube oscillators 129, 131–32, 135–36
carbon nanotubes
  see CNTs
  growth of 87, 96
  multiwalled 136, 170
  single-walled 72, 149–50, 169
carbon nanowalls 86–88, 97
carrier concentration 156, 179, 182–83, 186
catalyst 5–7, 11–12, 51, 79–80, 88, 91
catalyst materials 66
cathodoluminescence 232–34, 243, 247
CBED (convergent beam electron diffraction) 260–61
channel, single wavelength 215
channel direction 260, 268–70
channel surface 268, 270
chemical vapor deposition *see* CVD
chiralities 117–19, 127
CNT arrays 57–59, 62
CNT deposition on conducting substrates 65, 67, 69
CNT emitters 59–61, 63, 70
CNT thin films on flexible substrates 52–53, 55, 57
CNTs (carbon nanotubes) 9–10, 41, 50–52, 54–59, 61, 63–74, 86–87, 96–97, 112, 115, 118, 127–28, 130–31, 135–36, 148–50, 153–60, 163–70, 172
  ultra-short 154–55
cold cathodes 44, 46, 48, 71–72, 74
compositional mixing 251–53, 256
compressive strain 265–66, 268
concave lens 189, 191–92
conductance 2–3, 27, 138, 142, 145, 155, 157, 159, 172
conductivity 61, 137, 139, 141, 145, 148, 150, 158, 169, 173, 277
contact resistance 53–54, 104, 167
convergence angle 192, 194, 197
convergent beam electron diffraction see CBED
convex lens 189, 191–92, 194, 196–97

crystallinity 251–52, 256
cubic phase 237, 240
curvature 156, 234
CVD (chemical vapor deposition) 6–7, 20, 24, 35, 49, 72, 74–76, 78–88, 90, 92, 94–98, 139, 216, 223, 257

defect emissions 232
defect-induced electrical breakdown on individual MWCNTs 164, 166, 168, 170
defect sites 164, 168
defects 5, 61, 83, 91, 93, 118, 120, 122, 129–30, 132, 135–36, 150, 164–65, 167–69, 264
    local 169
degenerate regime 179, 184–86
dense wavelength division multiplexed *see* DWDM
deposition 5, 9, 13–14, 16, 20, 22–23, 66, 74, 96–97, 106, 138, 174, 252
devices, optical 9, 202
DHF (diluted hydrofluoric acid) 20, 207
diamond 22, 49, 72–74, 86, 97, 142, 145, 147, 149–50
diamond-like carbon *see* DLC
dichlorobenzene 105, 107–9
diffraction patterns 194, 261–63, 265
digital mirror device *see* DMD
diluted hydrofluoric acid *see* DHF
dispersion 53, 138–39, 149, 214
distortions 130–31, 133, 235, 239
DLC (diamond-like carbon) 49–50, 73, 260
DMD (digital mirror device) 189, 191, 193
DMD chip 189, 191–94, 196–97
donor 99–100, 102–4, 109, 112–13, 124–26
donor moiety 126
dopant concentration 234–35
doping 4, 30, 35, 232, 234, 243, 274, 276
dry etching 17, 20, 25, 216, 224
DWDM (dense wavelength division multiplexed) 213–14

e-beam lithography 11–12, 17, 23, 32, 190
EBL (electron beam lithography) 11, 18, 44, 57, 59, 62, 163, 165, 222
edge atoms 154, 156–57, 159
edge carbon atoms 153–54, 157
EFM (electrostatic force microscopy) 109
electric field 30–31, 43–44, 47–48, 54, 61, 72, 86
electrical breakdown 164–66, 168
electrodeposition 9–10, 13
electron beam lithography *see* EBL
electron diffraction pattern 237, 262
electron donor 100–1
electron emission 42–43, 45, 47, 49, 51, 55, 64, 72
electron emitters, state-of-the-art 44
electron field emission 73
electron microscopy *see* EM
electron mobility 274, 281–82
electron sources 71–73
electron transport properties 273–74, 281
electronic devices, molecular 117–18, 122
electronic structures 94, 117–19, 128, 154, 156–57, 159, 187
electrons 14, 42–43, 49, 52, 61, 73–74, 104–5, 110, 119, 126, 146, 156, 274–75, 278, 281–82
electrostatic force microscopy *see* EFM
EM (electron microscopy) 44, 47, 73, 235, 247
embedded-SiGe (eSiGe) structure 268–70
energy dissipation 111, 129–30, 133, 135–36
    modes of 133, 135
energy dissipation mechanisms 129–30, 135

energy gap 122–23, 127, 154, 156
etching 12, 14, 17, 20–23, 26, 31, 46, 159, 206, 229
  partial 207–8
  wet 17, 19, 22, 24

fast Fourier transform *see* FFT
FDTD (finite-difference-time-domain) 202–3
FE (field emission) 41–42, 49–51, 61, 66, 95, 233–36, 239–47
Fe-doped ZnO 233, 235–36, 240–41, 245–46
Fe doping 234–35, 239, 243, 245
Fermi–Dirac integral 179, 184–86
Fermi energy 184, 186, 278
few-layer graphene *see* FLG
FFT (fast Fourier transform) 276, 278–80
FIB (focus ion beam) 14–15, 222
field emission *see* FE
films
  flat 227–28
  rolled 54
finite-difference-time-domain *see* FDTD
FLG (few-layer graphene) 77, 89–92
focus ion beam *see* FIB
formation energy 117, 120–21, 127, 232
free spectral range *see* FSR
FSR (free spectral range) 214–15, 217–18

gas precursors 5, 7
gate oxidation 264–66
gold 9, 12–14, 174, 176
gold catalysts 11–13
graphene 50–51, 75–77, 79–80, 83–84, 86, 88–95, 97, 139, 150, 177, 180–81, 187
  fabricating 77, 79
  growing 86, 90–91
  growth of 79–80, 83–84, 87–88, 92–93, 95
  single-layer 77, 89–90
graphene-based micro-sensors 171–72, 174, 176
graphene films 79, 84, 97
graphene nanoribbon band structure 181, 183
graphene nanostructures 75–76, 78, 80, 82, 84, 86, 88, 90, 92, 94, 96, 98
graphene synthesis 77, 80, 97
growth 7, 10–12, 18, 67, 80–84, 86–89, 91, 93–95, 97, 118, 127, 206, 249, 257
  prismatic 234, 239
  template-based 8–9
growth of nanostructures 82, 84
Growth of Nanowires 4–5, 7, 9, 11, 13

hardmask 206–7
high-resolution transmission electron microscopes *see* HRTEM
highest occupied molecular orbital *see* HOMO
hole diameter 221, 224–26, 228
holes 104–5, 110–11, 164, 184, 221, 223, 225, 228
HOMO (highest occupied molecular orbital) 100, 104, 117, 125, 156
horizontal waveguide 203, 207–9
hot electrons 61, 74
HRTEM (high-resolution transmission electron microscopes) 67, 233, 237–38, , 260–61
hybrid approaches 4, 10–11, 13, 31
hydrocarbons 79, 87, 136
hydrogen 87, 97, 153, 155, 157, 159
hydrogenation 153, 157

ICP (inductively coupled plasma) 84, 216
indium tin oxide *see* ITO
inductively coupled plasma *see* ICP
ITO (indium tin oxide) 101, 138, 194

junctions
interfacial 118, 121–22
nanotube 128
three-layer 122, 125
tube–tube 139, 146
zigzag
armchair 121–23, 127

Kelvin probe force microscopy *see* KPFM
kinetic energy 46, 82–83, 132, 134
KPFM (Kelvin probe force microscopy) 100–1, 111, 115

LaB_6 emitters 44–46
laser interference patterning *see* LIP
laser spike annealing *see* LSA
line-scan strain measurement 265–67
LIP (laser interference patterning) 15
lithographically patterned NW electrodeposition *see* LPNE
local oxidation 18, 20
low-pressure chemical vapor deposition *see* LPCVD
lowest unoccupied molecular orbital *see* LUMO
LPCVD (low-pressure chemical vapor deposition) 7, 20, 251–53
LPNE (lithographically patterned NW electrodeposition) 13–14
LSA (laser spike annealing) 261, 264–67, 270
luminescence 243, 246
LUMO (lowest unoccupied molecular orbital) 100, 104, 117, 125–26, 154, 156

materials
acceptor 100, 103–4
crystalline 5, 22
luminescent 232
MBE (molecular beam epitaxy) 6, 12–13, 35, 273–74
mechanical exfoliation 76–77
metallic electrodes 65, 154–55
metallic NWs 2, 27–28
metals 6, 9, 13, 32, 35, 42, 44, 53, 79–80, 151, 165–66, 172–73, 222, 228, 268
microheaters 216–17
microplasma 95–96
microstructures 22, 235, 239, 243
molecular beam epitaxy *see* MBE
molecular devices 153–54
monolayer graphene 94–95
multi-beam interference 190, 192
MWCNTs 52, 149, 163, 165–66

nano-beam electron diffraction *see* NBED
nano-imprint lithography *see* NIL
nanodevices 22
nanoelectronics 1–2, 24, 169
nanohole arrays 222–23, 229–30
nanoimprint lithography 17, 37–38, 57
nanoparticles 100, 148, 171–72, 232
nanosheets, vertically aligned graphene 87
nanosphere 12, 224
nanosphere lithography technique 222, 229
nanostructures 17, 26, 37, 76, 82, 84, 87, 96, 112, 115, 172–73, 222–23, 229
complex 8
nanotubes 36, 50, 71, 73–74, 76, 84, 86, 96, 118–20, 122, 124, 127, 149–50, 170–73, 175–76
nanowires 1–2, 4–14, 16, 18, 20, 22, 24, 26, 28, 30, 32, 34, 36–40, 55, 74
NBED (nano-beam electron diffraction) 260–62, 265–66, 268, 270
nickel 14, 89
NIL (nano-imprint lithography) 17–18, 222
nitride stress liner *see* NSL
non-degenerate regime 179, 185–86
NSL (nitride stress liner) 260
nucleation 5–7, 83, 96

optical communication network 213–14, 216, 218–20
organic solar cells 99–100, 102, 104–6, 108, 110–12, 114
oscillation, quantum size 118, 122, 127
oscillators 130–33, 136
  defect-free 131–33, 135
  defected 129–31, 133, 135

pair defects 117, 119–20, 122–23
PCE (power conversion efficiency) 103, 106
PCs (photonic crystals) 189–90, 197–98
PDLC (polymer-dispersed liquid crystal) 189–90, 199
PDLC materials 190, 195, 197
PECVD (plasma enhanced chemical vapor deposition) 24, 76, 81–82, 88–92, 97, 216, 223
periodic field emitters 58–59, 61, 63
persistent photoconductivity 273–74, 276–78, 280–82
photonic crystals *see* PCs
photoresist 14–15, 190, 197, 206
planes, crystalline 18, 23–25, 33
plasma 14, 24, 51, 57, 76, 81–93, 95–97, 262
  microwave 96–97
plasma enhanced chemical vapor deposition *see* PECVD
plasma environment 87–88
plasma heating 82–83, 96
plasmonic nanostructures 222
plasmonic resonance wavelength 221–23, 225–26, 228–29
polarization 201–2, 205, 209–10
polarization rotator 202–5, 207–10, 212
  waveguide-based 201–2, 210
polymer 33, 38, 99–101, 104–6, 108, 113, 214
  conjugated 73, 101–2, 104
polymer composites 99–100, 102, 104, 106, 108, 110, 112, 114
polymer-dispersed liquid crystal *see* PDLC
polymer solar cells *see* PSCs
porosity 234–35, 243
power conversion efficiency *see* PCE
PSCs (polymer solar cells) 100–2, 104, 107–8, 112–14

quantum conductance 154, 157–59
quantum mobilities 278, 281–82

reactive ion etching *see* RIE
reconfigurable interference lithography system *see* RILS
reflective spectrum 224, 226
resistivity 2, 276–77
resonance, plasmonic 224–25, 228
resonance wavelength 228–29
resonators 211, 220
RIE (reactive ion etching) 20, 216, 223
RILS (reconfigurable interference lithography system) 189, 191, 194–95, 197

scanning electron microscopy *see* SEM
Schottky emitters 44, 46–47
SdH effect 274–75
SdH oscillations 276–77
SEM (scanning electron microscopy) 13, 30, 65, 67, 140, 168, 216, 224, 233–34, 243
sensors 1, 27, 39–40, 173–76
shallow trench isolation *see* STI
Si (silicon) 14, 18, 22, 32–33, 36, 38–39, 47, 59, 72, 77, 96, 201–3, 207, 210, 212, 214, 221, 222–23, 228–30, 252–56, 265–66
Si defects 264–66, 270
Si dots 252, 254–56
Si nanowires 35, 39–40
Si waveguides 202
silicon *see* Si
single-wall carbon nanotubes *see* SWCNTs
SMT (stress memorization technique) 260

SPT (stress proximity technique) 260
stainless steel mesh 66–68
STI (shallow trench isolation) 264, 266
Stone–Wales (SW) defects 130–32, 135
stress memorization technique *see* SMT
stress proximity technique *see* SPT
structures
  carbon-based 86
  filamentary 111–12
  three-layer 122, 125
SW defects, sp^3-hybridized 129–30
SWCNT-based TCFs 137–48
SWCNTs (single-wall carbon nanotubes) 117–18, 137–39, 142, 144–46, 155, 164, 180–81
synthesizing graphene 76–77, 79, 81, 90

TEM techniques 77, 260–61
tetra methyl ammonium hydroxide *see* TMAH
thermal annealing 13, 105–6
thermal interference 214–15, 218–19
TMAH (tetra methyl ammonium hydroxide) 23, 25
transistors 61, 187, 271, 274
transmittance 137–40, 146–48
transport, electronic 154
transport mobility 278–79
tubes, armchair 117, 119
tuning 6, 112, 214, 221–22, 226–27
tuning of plasmonic resonance wavelength 223, 229

vapor–liquid–solid mechanism 6–7, 11
virtual source size 46, 48, 69

water, de-ionized 53, 62, 67
waveguide structures 206, 209, 215
waveguide surface 207–8
wavelength 15, 77, 107, 111, 147, 204, 209, 213–15, 218, 221, 224–25, 229, 262
  drop 217–18
  extinction 222
wavelength channels 214
wavelength selective switches *see* WSS
wavelength tunability 214
wavelength tuning 218, 221, 227
WSS (wavelength selective switches) 213–15, 218–19

ZnO 8, 32, 55, 63, 232, 235–36, 239–40, 243–44
  nanostructure 232–33
ZnO NWs 11, 56, 62–63